Collins

Student Support Materials for AQA

A2 Physics

Unit 5: Nuclear and Thermal Physics and Option Units

Authors: Dave Kelly

William Collins's dream of knowledge for all began with the publication of his first book in 1819. A self-educated mill worker, he not only enriched millions of lives, but also founded a flourishing publishing house. Today, staying true to this spirit, Collins books are packed with inspiration, innovation and practical expertise. They place you at the centre of a world of possibility and give you exactly what you need to explore it.

Collins. Freedom to teach.

Published by Collins
An imprint of HarperCollinsPublishers
77-85 Fulham Palace Road
Hammersmith
London
W6 8JB

> Browse the complete Collins catalogue at
> **www.collinseducation.com**

10 9 8 7 6 5 4 3

ISBN   978-0-00-734386-7

Dave Kelly and Ron Holt assert their moral rights to be identified as the authors of this work.

British Library Cataloguing in Publication Data. A Catalogue record for this publication is available from the British Library.

Thanks to John Avison for his contribution to the previous editions.

Commissioned and Project Managed by Letitia Luff
Edited by Jane Roth and Geoff Amor
Proofread by Jane Roth, Anne Trevillion and Tony Clappison
Typesetting by Hedgehog Publishing
Cover design by Angela English
Index by Jane Henley
Production by Leonie Kellman
Printed and bound by Printing Express, Hong Kong

Acknowledgements

The publishers wish to thank the following for permission to reproduce photographs. Every effort has been made to trace copyright holders and to obtain their permission for the use of copyright material. The publishers will gladly receive any information enabling them to rectify any error or omission at the first opportunity.

Page 91, Fig 26: Caltech; page 95, Fig 28: NASA; page 212, Fig 18: Andreas Frank

MIX
Paper from responsible sources
FSC C007454

Contents

Evidence for the nucleus

Some early attempts to investigate matter at an atomic level involved firing **alpha particles** (see page 5) at matter and observing how they were scattered. In Ernest Rutherford's laboratories at Manchester University in 1909, Geiger and Marsden studied the scattering of alpha particles as they passed through a thin piece of gold foil (Fig 1).

Fig 1
Rutherford scattering apparatus

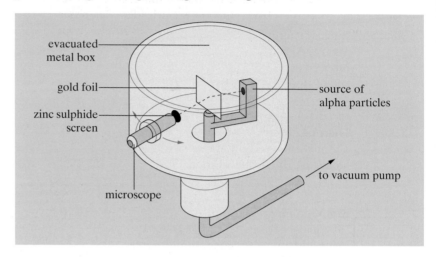

To detect the scattered alpha particles, Geiger and Marsden observed a scintillator (a zinc sulphide screen that emitted light when an alpha particle struck it). The vast majority of alpha particles were deflected from their path by very small angles, but on rare occasions an alpha particle would bounce back towards the source. The alpha particles were travelling with so much energy that they could not be bounced back by a gold atom that was a diffuse positively charged cloud embedded with electrons. Rutherford concluded that the positive charge, and almost all the mass of the atom, must be concentrated in one small volume – the nucleus. Most of the alpha particles passed through the foil without deflection, or with very small deflection, because they did not pass near enough to the charged nuclei. Very occasionally an alpha particle would pass close to a nucleus and experience a strong electrostatic force of repulsion (Fig 2).

Essential Notes

Before this experiment, the atom was thought of as a positively charged, uniformly dense mass with negatively charged electrons embedded in it. This was J. J. Thomson's 'plum pudding' model.

Fig 2
Rutherford scattering

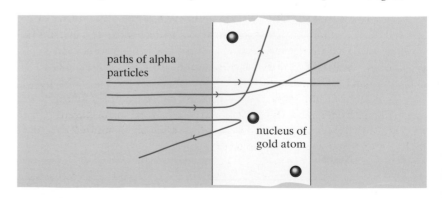

α, β and γ radiation

The phenomenon of radioactivity had been discovered by Henri Becquerel in Paris in 1897. Becquerel developed some photographic film that had been kept near to a sample of uranium salts. Although the film had been kept in light-tight wrapping paper, it was blackened, just as if it had been exposed to light. Becquerel concluded that the uranium compound was emitting some invisible radiation that could penetrate paper and darken photographic film. The invisible radiation obviously carried a significant amount of energy. Further experiments showed that it could cause **ionisation** in air molecules. Marie and Pierre Curie identified other elements, such as radium, that emitted radiation with similar properties.

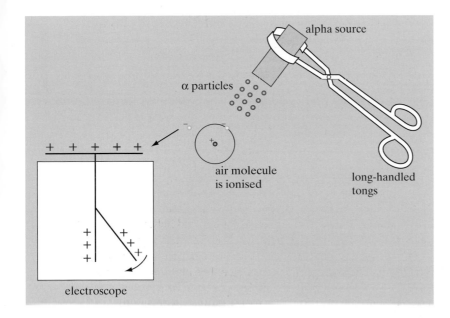

electroscope

Fig 3
Nuclear radiation is often referred to as 'ionising' radiation. The radiation has enough energy to remove electrons from atoms. These free electrons can then neutralise a positively charged electroscope

Experiments by Becquerel, Rutherford and others identified three distinct types of ionising radiation, which are now referred to as **alpha** (α), **beta** (β) and **gamma** (γ) radiation.

Alpha radiation

Alpha radiation is the least penetrating of these radiations. It is made up of particles, which are positively charged and relatively massive. Rutherford showed that an alpha particle is identical to a helium nucleus. Each alpha particle is a tightly knit group of two neutrons and two protons, held together by the strong nuclear force. An alpha particle can be fired out from an unstable nucleus, usually a large nucleus like radium. The alpha particle is emitted with an energy of up to 10 MeV. The energy for this decay comes from a mass difference between the original, or **parent nucleus**, and the decay product, or **daughter nucleus**. Bismuth-212 is an alpha emitter that decays to thallium:

$$^{212}_{83}\text{Bi} \rightarrow\ ^{208}_{81}\text{Tl} + ^{4}_{2}\text{He}$$

Alpha particles have only a short range in air, typically 5 cm, and are stopped by a thin sheet of material such as paper or skin. Alpha radiation is intensely ionising, creating tens of thousands of ion pairs per cm in air.

Alpha radiation is not a major risk to health, provided that the emitter is *outside* the body, since its limited penetration means that all the energy is dissipated in the outer layer of skin. However, alpha emitters are very damaging when ingested, since all the energy is deposited in a small volume. Radon is an alpha-emitting radioactive gas that can accumulate in buildings, particularly in parts of the country with granite rocks, like Cornwall. Radon has been identified as increasing the risk of lung cancer.

Fig 4
Alpha radiation inside the body is particularly damaging to DNA, since it is likely to cause multiple breaks in the double helix that are difficult for the cell to repair. It may also ionise other molecules in a cell, which may cause chemical damage to the DNA

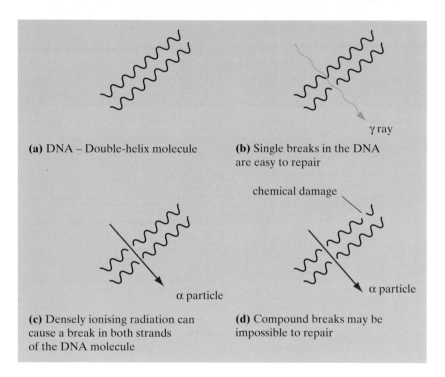

(a) DNA – Double-helix molecule

(b) Single breaks in the DNA are easy to repair

(c) Densely ionising radiation can cause a break in both strands of the DNA molecule

(d) Compound breaks may be impossible to repair

Alpha radiation is used in radiotherapy in the form of radium needles, to treat cancer. Alpha radiation from americium-241 is used in smoke detectors to cause an ionisation current in a small chamber. Smoke particles are larger and less mobile than air molecules, so when smoke particles enter the chamber they reduce this ionisation current. The drop in current causes an alarm to be sounded.

Beta radiation

Beta particles are very fast moving electrons that are emitted from the nucleus of some **radioisotopes**. This happens when a neutron decays into a proton and an electron. The proton remains inside the nucleus and the electron is emitted at close to the speed of light. An antineutrino, $\bar{\nu}_e$, is also emitted.

Strontium-90 is a beta emitter that is produced in nuclear reactors as one of the waste products of the nuclear fission of uranium. The equation describing its decay is

$$^{90}_{38}\text{Sr} \rightarrow \,^{90}_{39}\text{Y} + \,^{0}_{-1}\text{e} + \bar{\nu}_e$$

Essential Notes

Remember that isotopes are atoms of the same element, with the same number of protons in their nuclei but a different number of neutrons. **Radioisotopes** are radioactive elements whose nuclei have an unstable arrangement of nucleons (see page 17).

Beta particles have a longer range in air than alpha particles, typically 2 to 3 m. They are rather more penetrating and can travel through thin sheets of absorber such as plastic or paper. Beta radiation is less densely ionising than alpha radiation. Because beta particles are negatively charged, they are deflected by electric and magnetic fields.

Beta radiation presents a health hazard to humans when the source is outside the body, though the penetration is quite small. Most of the radiation damage is to the skin and surface tissues. A beta radiation source inside the body causes less local damage than an alpha source because its energy is dissipated over a larger volume. However, like all ionising radiation, a significant dose of beta radiation will damage DNA and lead to cell death, mutations or cancer.

Beta radiation is used to monitor the thickness of paper in a paper mill. If the paper is too thick, more beta radiation will be absorbed and the count rate will drop. An automatic adjustment can be made to the production process.

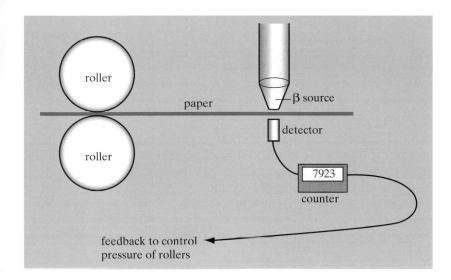

Fig 5
Beta radiation is used to monitor paper thickness

Gamma radiation

Gamma radiation is very high-frequency electromagnetic radiation that is emitted from the nucleus of some radioisotopes. This emission does not change the nucleus to that of another isotope, but the nucleus does reduce its energy. An example of a gamma emitter is cobalt-60, which is often used in radiotherapy. The cobalt-60 nucleus is in a higher energy state than normal and it decays to the ground state, emitting a gamma ray photon in the process:

$$^{60}_{27}\text{Co} \rightarrow {}^{60}_{27}\text{Co} + \gamma$$

Gamma rays are not charged and so cannot be deflected by electric or magnetic fields. They interact with matter less strongly than alpha or beta particles and are much less ionising. As a result of this, gamma radiation is very penetrating and can easily pass through thin sheets of metal. In fact thick sheets of steel or lead are used to shield against gamma radiation. Even so, the shielding just reduces the intensity of the gamma radiation and does not completely absorb it. See Fig 6.

Fig 6
Exponential absorption of gamma radiation

For gamma rays of a given energy it takes a fixed thickness of shielding to reduce the intensity by half

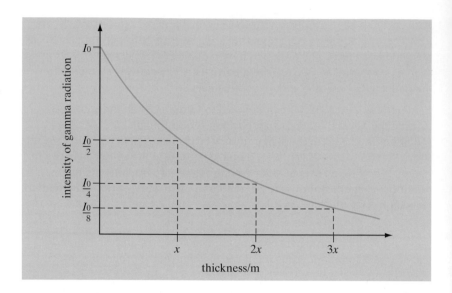

Gamma rays present a danger to human health even when the source is some distance from the body. A large dose of external gamma radiation could well have the same biological impact as a low dose of internal alpha particles. The best way of protecting yourself from gamma radiation is to keep well away from the source, since the intensity decreases with distance.

Gamma-emitting radioisotopes are often used as **tracers**. A small amount of radioactive gas can be introduced into a pipeline. If the gamma intensity above the ground is then measured, it is possible to find a leak. Gamma radiation is used in industrial radiography to produce 'shadow' pictures, in the same way that X-rays are used in medicine. A gamma-emitting source can be placed into a welded pipeline with photographic film around the outside. Any cracks in the welding will be shown as an overexposed line on the film.

The inverse square law

The **intensity**, I, of gamma radiation at any point is the power that flows through an area of one square metre. Intensity is measured in watts per square metre, $W\,m^{-2}$. The gamma radiation from a small source can be considered to be the same in all directions (isotropic) and therefore the energy spreads out over the surface of a sphere. The intensity at a distance x away from a source of constant intensity is given by the inverse square law (see Fig 7):

$$I = \frac{k}{x^2}$$

where k is a constant.

This has important implications for safe handling of a gamma source. The inverse square law means that doubling the distance from the source decreases the radiation dose to one-quarter. It is important to use long-handled tongs to manipulate the source. Gamma sources should be stored well away from people.

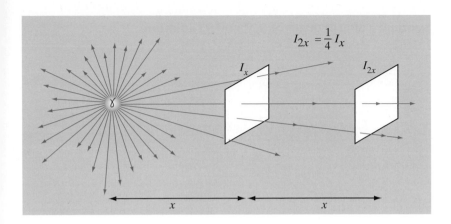

Fig 7
Inverse square law for gamma radiation

Any radiation that emanates from a point source, such as light from a small bulb, will follow the inverse square law

Essential Notes

The intensity of alpha and beta radiation does not follow the inverse square law, because these particles are absorbed and deflected by molecules of air to a much greater extent than gamma radiation. In a vacuum the law would apply equally to all three of these forms of radiation.

Example

At a distance of 20 cm from a gamma source the intensity is $1.6 \, \mu W \, m^{-2}$. At what distance will the intensity drop to $0.1 \, \mu W \, m^{-2}$?

Answer

Use the equation

$$I = \frac{k}{x^2}$$

At a distance of 0.20 m

$$1.6 \times 10^{-6} = \frac{k}{0.2^2}$$

Apply the equation again at a distance x

$$0.1 \times 10^{-6} = \frac{k}{x^2}$$

If we divide these equations,

$$\frac{1.6}{0.1} = \frac{x^2}{0.22}$$

$$x^2 = 0.64$$

$$x = 0.80 \text{ m}$$

Summary of properties of α, β and γ radiation

Table 1

Radiation	Nature	Penetrating power	Range in air	Ionising effect	Behaviour in electric and magnetic fields
Alpha	Two protons and two neutrons (helium nucleus)	Easily stopped, e.g. by a sheet of paper or the outer layer of (dead) skin cells	A few cm	Intensely ionising: an alpha particle will cause about 10^4 to 10^5 ion pairs per cm in air	Positively charged, so deflected by electric and magnetic fields; but relatively massive, so deflected less than beta particles
Beta	An electron	Stopped by thin (a few mm)* metal sheet	Several metres	Less intensely ionising than alpha: a beta particle will cause about 1000 ion pairs per cm*	Negatively charged, so deflected in opposite direction to alpha; deflected more than alpha as the mass is much less
Gamma	High-frequency electro-magnetic radiation	Reduced in intensity by $\frac{1}{2}$ by about 5 cm of concrete or 1 cm of lead*	10 to hundreds of metres*	Weakly ionising: about 10 ion pairs per cm*	Not charged, so undeviated by a magnetic or an electric field

*The actual value depends on the initial energy with which the radiation is emitted.

Example

A sample of radioactive material is discovered. Explain how you would determine whether it is emitting alpha, beta or gamma radiation.

Answer

The different penetrating powers of the different radiations can be used to identify the radiation emanating from the sample. Different barriers are put between the source of radiation and the detector (GM tube plus counter).

- If the count rate drops when the barrier is paper, there must be some alpha radiation present.
- If the count rate drops when aluminium is used as the barrier, then beta radiation is present.

- If there is a non-zero count rate when thick aluminium is used, then there is some gamma radiation present.

The background count rate (see page 12) should be taken and subtracted from each reading.

Alternatively, a magnetic field could be used, although a strong magnetic field is needed to deflect alpha particles significantly.

Background radiation

We live in a radioactive world. Every day we are exposed to nuclear radiation from the air that we breathe and the rocks that we walk on. We are also exposed to a small radiation dose due to medical and industrial procedures, though the largest fraction (around 87%) of the dose is from natural sources. See Fig 8.

The major sources of background radiation dose are:

- **Air.** This dose is mainly from radon and thoron, radioactive gases that are part of long-lived decay series.

- **Rocks and buildings.** Some rocks, particularly granite, contain uranium-238 or thorium-232. These two isotopes have long half-lives and decay to other radioactive products. The radiation dose from this source depends on where you live.

- **Cosmic rays.** The Earth is continually being bombarded with high-energy particles and gamma rays from space, mainly from the Sun but also from other sources outside the Solar System. Fortunately the atmosphere screens us from the worst effects of this radiation.

- **Food and drink.** Radioactive isotopes dissolved in water are taken up by plants, and then animals. Some foods are naturally more radioactive than others because they tend to concentrate radioactive isotopes. People are radioactive mainly due to the isotope potassium-40, which is concentrated in muscle.

- **Medical** procedures also contribute to our radiation dose. Most of this is due to diagnostic X-rays, but nuclear medicine techniques that use radioisotopes are becoming more common.

- **Miscellaneous.** Artificial background radiation also comes from industrial techniques, mining wastes, nuclear power, and even from the fall-out from nuclear weapons tests that were carried out in the 1950s and 1960s.

Essential Notes

Most naturally occurring radioisotopes are part of a decay series – a chain of radioactive decays in which a radioactive element decays to another radioactive element, and so on until a stable isotope is attained. The intermediate radioisotopes may have half-lives from millions of years to just minutes.

Fig 8
Sources of background radiation dose

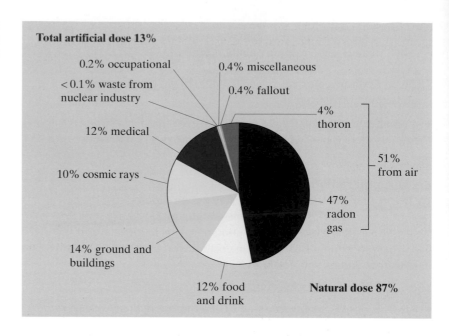

Total artificial dose 13%

0.2% occupational

<0.1% waste from nuclear industry

0.4% miscellaneous

0.4% fallout

4% thoron

12% medical

51% from air

10% cosmic rays

47% radon gas

14% ground and buildings

12% food and drink

Natural dose 87%

The **background radiation count rate** has to be taken into account when measurements of radioactivity are undertaken in the laboratory.

Essential Notes

A Geiger counter detects the ionisation caused by radiation. Although a Geiger counter is quite effective at detecting beta radiation, it is much less efficient at counting alpha or gamma radiation, and so will significantly underestimate the activity of alpha or gamma emitters.

Activity

The **activity** of a radioactive source measures the number of decays that occur, on average, every second. Activity is measured in **becquerel**, Bq. One becquerel is one disintegration per second. The activity of a source is measured with a Geiger counter.

The inverse square law for gamma radiation can be demonstrated using a Geiger counter to measure the count rate. The count rate has to be corrected by subtracting the background count from each reading.

In a school laboratory a Geiger-Müller tube, known as a GM tube, and an electronic counter, known as a digicounter, are used to measure count rate.

Fig 9
Experimental arrangement for demonstrating the inverse square law

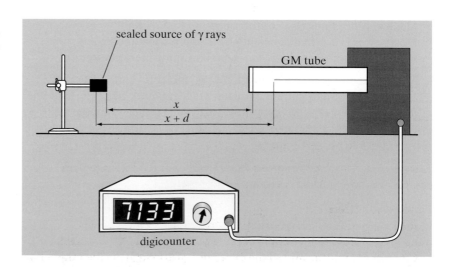

sealed source of γ rays

GM tube

x

$x + d$

digicounter

The **corrected count rate**, C, is recorded at a number of distances, x, from the gamma source. There is a systematic error in the measurement of x, because it is difficult to measure exactly where the gamma rays are emitted from and where they are detected. This uncertainty adds an average distance d to the distance (see Fig 9). If the inverse square law is correct then

$$C \propto \frac{1}{(x + d)^2}$$

So, introducing a constant k,

$$x = kC^{-1/2} - d$$

Compare this to the equation for a straight line, $y = mx + c$. A graph of distance, x, on the y-axis against $C^{-1/2}$ on the x-axis should yield a straight line with a gradient of k and an intercept of $-d$.

Radioactive decay

Radioactive decay is a *random* event. For a given sample of material, there is no way of predicting with any certainty which nucleus will decay, or when it will decay. All we can do is to give the probability that a nucleus will decay in a given time. Some radioisotopes decay relatively quickly, whereas some take thousands of years. For every radioisotope the probability that a nucleus will decay in a certain time is given by the **decay constant**, λ. In a large number of radioactive nuclei, N, the number of atoms that decay in a certain time, Δt, is ΔN. The probability that any particular nucleus will decay in time Δt is

$$\frac{-\Delta N/N}{\Delta t} = \lambda$$

This equation is usually written as

$$\frac{\Delta N}{\Delta t} = -\lambda N$$

or

$$\text{activity } A = -\frac{\Delta N}{\Delta t} = \lambda N$$

This equation says that the activity of a radioactive source, $\Delta N/\Delta t$, is proportional to the number of active nuclei present. The decay constant λ is the constant of proportionality, which is different for each radioisotope.

Radioactivity is also *spontaneous*. The rate of decay is not influenced by external factors, such as temperature or pressure.

Exponential decay

Consider a sample of a radioisotope. As time goes by the number of radioactive nuclei in the sample of material will reduce. At first there are a large number of nuclei and so the decay rate is also high. As the number of

Examiners' Notes

The negative sign in these equations indicates that the change in the number of atoms, ΔN, is a *decrease*.

Essential Notes

We can model the random nature of radioactive decay by thinking about dice. Each radioactive atom is like a single die. If we throw hundreds of dice at once, we cannot predict which die will show the number '6', but we can predict that roughly one in six will show a '6'. If the dice are truly random, the only factor that will affect the number of sixes is the total number of dice thrown.

Essential Notes

You met an example of exponential decay in Unit 4: the charge vs time graph for a discharging capacitor follows the same exponential law.

nuclei reduces, the decay rate also reduces. This kind of behaviour is known as **exponential decay**.

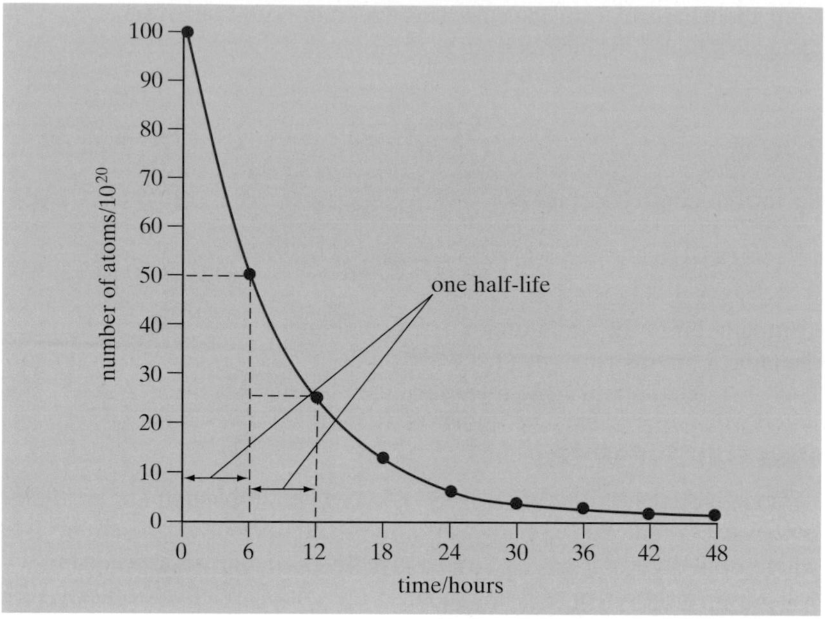

Fig 10
Exponential decay curve showing how the number of radioactive nuclei changes with time

Examiners' Notes

It is usual to use t in seconds and λ in units of s^{-1}. However sometimes it is more convenient to use time in years. λ must then have units of $(years)^{-1}$.

The equation that describes the graph in Fig 10 is

$$N = N_0 e^{-\lambda t}$$

where N_0 is the initial number of atoms present, i.e. the number at time $t = 0\,s$, λ is the decay constant, and t is the time.

The activity of a source, measured in becquerels, is also proportional to the number of active nuclei present: $A = \lambda N$. The activity of a source therefore follows the same exponential decay law:

$$A = A_0 e^{-\lambda t}$$

Example

Radioactive iodine-131 has a decay constant of $9.9 \times 10^{-7}\,s^{-1}$. A sample of iodine-131 has an initial mass of $0.10\,g$.

(a) Calculate the initial activity of the source.

(b) Calculate the activity after 30 hours.

Answer

(a) The activity, A or $\Delta N / \Delta t$, is given by $-\Delta N / \Delta t = \lambda N$.

The number of atoms present in a mole of iodine is 6.02×10^{23} (this is the Avogadro constant, see page 39).

A mole of iodine-131 would have a mass of 131 g. So 0.10 g represents $0.10/131 = 7.63 \times 10^{-4}$ of a mole, which is $7.63 \times 10^{-4} \times 6.02 \times 10^{23} = 4.60 \times 10^{20}$ atoms.

So the activity is

$$\lambda N = 9.9 \times 10^{-7} \times 4.60 \times 10^{20} = 4.55 \times 10^{14} \text{ Bq}$$

(b) The activity after 30 hours follows the rule $A = A_0\, e^{-\lambda t}$.

$$t = 30\,\text{h} = 30 \times 3600 = 1.08 \times 10^5\,\text{s}$$

So $A = 4.55 \times 10^{14} \times e^{-(9.9 \times 10^{-7} \times 1.08 \times 10^5)}$

$$= 4.09 \times 10^{14}\,\text{Bq}$$

Half-life

Every radioactive isotope has its own **half-life**. This is the time taken for the number of active nuclei in a sample of radioactive material to drop to half of the original value.

Essential Notes

1 Bq is equivalent to 1s^{-1}

Definition

The half-life is the time it takes for the number of active nuclei in a sample to drop to half of its original value.

The half-life is also the time it takes for the activity to drop to half of its original value. After two half-lives the activity will drop to $\frac{1}{4}$ of its original value and after three half lives the activity will only be $\frac{1}{8}$ of the original activity.

The half-life is linked to the decay constant, λ. An isotope with a small value of λ has nuclei with a low probability of decay, so the isotope has a long half-life.

The exact relationship can be deduced from the equation $N = N_0\, e^{-\lambda t}$. After one half-life N will equal $N_0/2$:

$$\frac{N_0}{2} = N_0 e^{-\lambda t} \quad \text{or} \quad \frac{1}{2} = e^{-\lambda t}$$

Taking logarithms of both sides of this equation gives

$$\ln\left(\frac{1}{2}\right) = -\lambda t$$

Since $\ln\left(\frac{1}{2}\right) = -\ln 2$

$$\text{half-life } T_{\frac{1}{2}} = \frac{\ln 2}{\lambda}$$

It is possible to plot a logarithmic–linear (see page 50) graph to find the half-life of an isotope. Suppose that we measure the activity, $A = -\Delta N/\Delta t$, at several different times, t.

Essential Notes

We have used the fact that the logarithm of a product is equal to the sum of the logs of the individual numbers,

$$\log(A \times B) = \log A + \log B$$

Since $A = A_0\, e^{-\lambda t}$, taking logarithms of both sides gives

$$\ln A = \ln (A_0\, e^{-\lambda t})$$

This becomes

$$\ln A = \ln A_0 + \ln (e^{-\lambda t})$$
$$\ln A = \ln A_0 - \lambda t$$

Compare this with the equation of a straight line, $y = c + mx$. If we plot ln A on the y-axis and t on the x-axis, we will get a straight line with a gradient of $-\lambda$. Since λ is the decay constant, we can then calculate the half-life from the gradient by using the relationship

$$T_{\frac{1}{2}} = \frac{\ln 2}{\lambda}$$

Storage of radioactive waste

Some of the isotopes created as fission products in nuclear reactors are highly radioactive. These isotopes have a high value of λ and a short half-life. The spent fuel rods are stored under water in large ponds for several months until the activity from these isotopes has reduced (see page 30). Some of the other products, like plutonium, are much longer lived. Plutonium has a half-life of 20 000 years and presents long-term storage problems.

Example

Strontium-90 is one of the radioactive isotopes that are produced as waste products in fission reactors. The activity of a small sample of strontium-90 was measured over a period of 10 years and the following results were recorded.

Time/years	Activity/MBq
0	200.0
1	195.1
2	190.3
3	185.7
4	181.1
5	176.7
6	172.4
7	168.2
8	164.1
9	160.1
10	156.1

Plot a suitable graph to calculate the decay constant, and hence the half-life, of strontium-90.

Answer

Since $\ln A = \ln A_0 - \lambda t$, we can plot $\ln A$ against the time, t. The gradient will give us the decay constant.

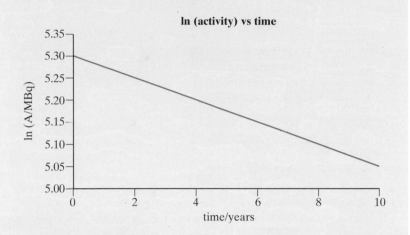

ln (activity) vs time

Fig 11
Using a logarithmic–linear graph to find half-life

The gradient is equal to $-0.025 \text{ years}^{-1}$, so the half-life is $(\ln 2)/0.025 = 27.7$ years.

Radiocarbon dating

The radioactivity of a carbon isotope, carbon-14, is used to date artefacts that are made from organic materials. Carbon-14 is formed in the upper atmosphere by the action of cosmic rays on nitrogen. When nitrogen absorbs a neutron it decays to carbon-14:

$$^{14}_{7}\text{N} + ^{1}_{0}\text{n} \rightarrow ^{14}_{6}\text{C} + ^{1}_{1}\text{H}$$

The carbon-14 forms part of the atmospheric carbon dioxide and it becomes assimilated into plants through photosynthesis. Every living plant contains a small, known, percentage of radioactive carbon. During a plant's life the level of radioactive carbon stays more or less constant, the plant takes in new carbon-14 to replace that which has decayed. However, as soon as the plant dies the level of radioactivity from carbon-14 begins to drop. Since we know that carbon-14 has a half-life of 5730 years, it is possible to estimate the age of the artefact from the activity.

Nuclear instability

Stable isotopes with low atomic numbers tend to have equal numbers of neutrons and protons in their nuclei. The nuclei are stable because the attraction of the strong nuclear force, which acts between all nucleons, is strong enough to balance the repulsive electrostatic force. This is not the case for isotopes with higher atomic numbers. Because these have more protons there is a greater electrostatic force pushing the nucleus apart. More neutrons are needed to 'glue' these nuclei together. Stable isotopes with large nuclei therefore have more neutrons than protons.

Essential Notes

The electrostatic force has a much greater range than the strong nuclear force, which is limited to around 10^{-15} m. Two protons on the opposite side of a large nucleus still repel each other due to the electrostatic force, but are out of range of the strong nuclear force.

17

The stable nuclei are shown in Fig 12. Nuclei that lie either above or below this curve tend to be unstable.

Fig 12
Graph of neutron number, N, against proton number, Z

Vertical lines represent isotopes. Isotopes which are further from the main curve tend to be more unstable

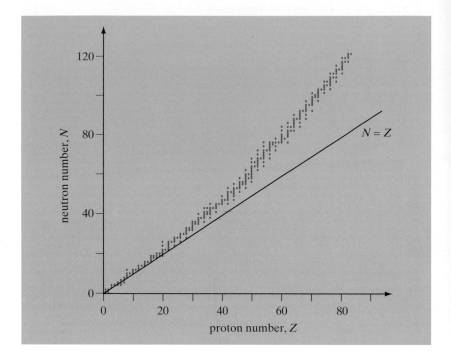

Possible decay modes of unstable nuclei

There are a number of ways in which unstable nuclei can decay.

Alpha (α) decay

Alpha emitters are proton-rich heavy isotopes ($Z > 60$). An alpha particle is a very stable combination of two protons and two neutrons, identical to a helium nucleus, that can form within a large nucleus. After many attempts, the alpha particle may escape from the nucleus.

Alpha decay reduces the proton number, Z, of the parent nucleus by 2, and reduces the nucleon number, A, by 4. The general equation for alpha decay is

$$^A_Z X \rightarrow {}^{A-4}_{Z-2} Y + {}^4_2 He$$

where X is the parent (alpha-emitting) isotope and Y is the daughter. A gamma ray is often also emitted (see page 7).

Beta minus (β^-) decay

Beta decay occurs when a neutron is transformed into a proton, with the emission of a high-speed electron and an antineutrino from the nucleus.

Beta decay leaves the nucleon number unchanged, since a neutron has been transformed into a proton. The proton number goes up by 1. The general equation for beta-minus decay is

$$^A_Z X \rightarrow {}^A_{Z+1} Y + {}^0_{-1} e + \nu_e$$

Essential Notes

This is a random process, though the probability of emission is greater if the alpha particle has more energy. In fact there is a link between alpha particle energy and the half-life of the isotope; the greater the alpha's energy, the shorter the half-life.

Beta-minus emitters are neutron-rich and lie above the curve of stable nuclei in Fig 10. A free neutron will decay into a proton and an electron (beta-minus) with a half-life of around 11 minutes. This decay is energetically possible because the mass of a neutron is slightly greater than the combined mass of the proton and electron.

Beta plus (β^+) decay

Beta plus decay occurs when a proton is transformed into a neutron, emitting a high-speed positron and a neutrino from the nucleus. The positron is the antimatter version of the electron. It has exactly the same mass as the electron. The positron's charge is $+1.6 \times 10^{-19}\,$C, equal in size but opposite in sign to the charge of the electron.

Beta-plus decay leaves the nucleon number unchanged, but the proton number goes down by one. The general equation for beta-plus decay is

$$^A_Z X \rightarrow\, ^A_{Z-1} Y +\, ^0_{+1} e + \nu_e$$

Beta-plus emitters are situated below the N vs Z curve for stable nuclei (Fig 10).

Electron capture

It is possible for the nucleus of an atom to capture one of the atom's orbiting electrons. When this happens a proton in the nucleus absorbs the electron and becomes a neutron. This is how the isotope beryllium-4 decays:

$$^7_4 Be +\, ^0_{-1} e \rightarrow\, ^7_3 Li + \nu_e$$

Nucleon emission

It is possible, though rare, for an unstable isotope to emit a nucleon. A nucleus that is proton-rich, such as lithium-5, can emit a proton:

$$^5_3 Li \rightarrow\, ^4_2 He +\, ^1_1 p$$

Similarly helium-5, which is neutron-rich, emits a neutron:

$$^5_2 He \rightarrow\, ^4_2 He +\, ^1_0 n$$

Existence of nuclear excited states

Alpha emission is often accompanied by gamma-ray emission. After the alpha emission the daughter nucleus is left in an excited state. At some point this excited nucleus will decay by the emission of a gamma ray. The energy available for the decay is therefore shared between the alpha particle and the gamma ray.

The gamma rays emitted following alpha emission have a line spectrum that reflects the energy levels in the daughter nucleus.

Essential Notes

The ν_e is an antineutrino. Its existence makes sure that the laws of energy and momentum conservation are observed (see Unit 1).

Essential Notes

The decay of a proton into a neutron and a positron only happens inside certain unstable nuclei. A free proton is a very stable particle.

Essential Notes

The daughter nucleus also has some kinetic energy due to its recoil after the decay. However, the mass of the daughter tends to be much larger than the alpha particle. The daughter moves at a much lower velocity than the alpha and so it carries much less kinetic energy.

Fig 13
Bismuth-212 decays to thallium-206 by alpha emission. The decay can leave thallium in an excited state. The thallium later emits a gamma ray to return to the ground state

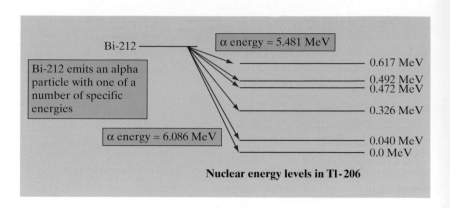

Nuclear energy levels in Tl-206

Examiners' Notes

A radioisotope used for medical diagnosis needs to have a half-life long enough to enable the investigation to be carried out, but not too long or the activity will be low and the patient will remain radioactive for too long. Gamma emitters are used because the radiation is not absorbed by the body.

Sometimes the gamma emission can be delayed by a significant time. The isotope is then marked with an 'm' to indicate that it is a **metastable state**. Technetium-99m is such an isotope. This is created by the decay of molybdenum-99, and has a half-life of 6 hours. The half-life of technetium-99m, and the fact that it is a gamma emitter, makes it ideal for use in hospitals as a tracer and for imaging.

Nuclear radius

Closest approach of an alpha particle

In an alpha scattering experiment such as Rutherford's (see page 4), when an alpha particle approaches a gold nucleus on a direct collision course it slows down and stops for an instant before rebounding. The alpha particle's kinetic energy is transferred to potential energy as the positively charged alpha particle does work against the electrostatic repulsion of the positively charged gold nucleus.

At the distance of closest approach, all the kinetic energy has been transferred to potential energy:

$$E_k = E_p$$

For a spherical charge, Q, the potential energy is

$$E_p = qV = \frac{Qq}{4\pi\varepsilon_0 r}$$

where V is the electric potential a distance r from the nucleus of charge Q (see Unit 4).

For a 5 MeV alpha particle approaching a gold nucleus of atomic number 79, this becomes

$$(5\times10^6 \times 1.6\times10^{-19})\,\mathrm{J} = \frac{(79\times1.6\times10^{-19})\mathrm{C}\times(2\times1.6\times10^{-19})\mathrm{C}}{4\pi\times8.85\times10^{-12}\,\mathrm{Fm^{-1}}\times r}$$

giving a value for r of 4.55×10^{-14} m. This gives an upper limit for the radius of the gold nucleus. Modern measurements give a value of 6.5×10^{-15} m, or 6.5 fm, **femtometres**.

Electron diffraction

All particles have a wave-like nature. The wavelength, λ, of an electron can be calculated from the momentum, p, of the electron using de Broglie's relation:

$$\lambda = \frac{h}{p}$$

where h is the Planck constant.

High-energy electrons can have a wavelength that is small enough to diffract around the nucleus. We can estimate the nuclear radius from these diffraction patterns.

Electrons can be accelerated by allowing them to pass through a potential difference, V. The larger the potential difference, the more energy the electron will gain and the smaller its wavelength will be.

The energy gained by the electron is $E = eV$, where $e = 1.6 \times 10^{-19}$ C is the charge on the electron. This energy will be equal to the final kinetic energy of the electron, $E_k = \frac{1}{2}mv^2$:

$$eV = \frac{1}{2}mv^2$$

This gives $m^2v^2 = 2\,meV$

So

$$p = mv = \sqrt{2\,meV}$$

The **de Broglie wavelength** is therefore

$$\lambda = \frac{h}{\sqrt{2\,meV}}$$

A potential difference of 100 V will give rise to electrons with a wavelength of 1.23×10^{-10} m.

By using higher voltages we can diffract an electron beam around the nucleus of an atom. The electrons are fired at a thin slice of material to produce a diffraction pattern.

The angle at which the first minimum of such a diffraction pattern appears (see Fig 14) is given by

$$\sin\theta = \frac{0.61\,\lambda}{R}$$

where R is the radius of the obstacle.

Electrons with an energy of 125 GeV have a de Broglie wavelength of 3.47×10^{-15} m, or 3.47 fm. An angle θ of 38° for oxygen nuclei gives a value for the radius of an oxygen nucleus of:

$$R = \frac{0.61 \times 3.46 \times 10^{-15} \text{ m}}{\sin 38°}$$
$$= 3.4 \times 10^{-15} \text{ m}$$

Essential Notes

Wave–particle duality was discussed in Unit 1.

Essential Notes

Compare this electron wavelength to the shortest wavelength of visible light, which is 400 nm or 4×10^{-7} m. The wavelength of the electron is more than 1000 times shorter.

Fig 14
Electron diffraction around a nucleus

The first diffraction minimum due to an oxygen nucleus is at about 38°

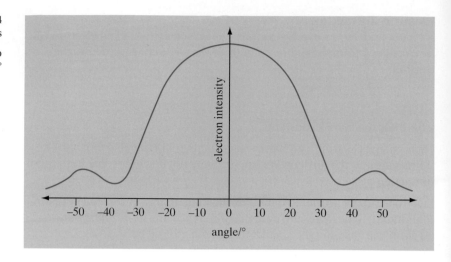

Diffraction experiments using different elements have shown that there is a link between the nucleon number, A, of a nucleus and its radius, R:

$$R = r_0 A^{1/3}$$

where r_0 is a constant representing the radius of a single nucleon.

Example

The following data are taken from an electron diffraction experiment.

Nucleon number, A	Radius R/fm
12	3.04
16	3.41
28	3.92
40	4.54
51	4.63

Plot a graph to verify that the nuclear radius, R, depends on the atomic number, A, according to this equation $R = r_0 A^{1/3}$.

Use your graph to find the value of r_0.

Answer

Because we are testing the relationship $R = r_0 A^{1/3}$, we need to plot the radius, R, against $A^{1/3}$ to get a straight line. The gradient will be the value of r_0.

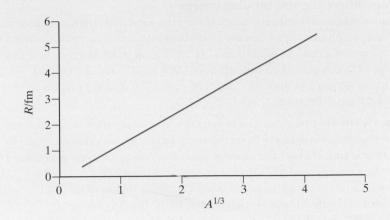

Fig 15

The gradient of 1.003 fm is the value of r_0.

Nuclear density

We can use the expression for the nuclear radius to investigate a value for the density of a nucleus.

The mass of a nucleus is approximately equal to $A \times m_n$, where A is the nucleon number and m_n is the mass of one nucleon. (There is a small difference between the mass of a proton and the mass of a neutron, but we will neglect that here.)

The volume of the nucleus (treating it as a sphere of radius R) is $V = \frac{4}{3}\pi R^3$.

Using $R = r_0 A^{1/3}$,

$$V = \frac{4}{3}\pi (r_0 A^{\frac{1}{3}})^3$$

$$= \frac{4}{3}\pi r_0^3 A$$

$$= A \times V_n$$

where V_n = the volume of one nucleon.

This gives the density as

$$\frac{A \times m_n}{A \times V_n} = \frac{m_n}{V_n}$$

This corresponds to the density of a single nucleon. The density of nuclear matter is therefore independent of which particular isotope we are considering. Nuclear matter has a density of approximately $2 \times 10^{17}\,\text{kg m}^{-3}$.

Essential Notes

Density, $\rho = \dfrac{\text{mass}}{\text{volume}}$

Essential Notes

This is an enormous density. If a table-tennis ball was made out of nuclear matter it would have a mass of about 8 million tonnes. If you squashed the Earth until it reached the density of nuclear matter, it would have a radius of only 200 m.

3.5.2 Nuclear energy

Mass and energy

Mass difference and binding energy

The total mass of a nucleus is not simply the total mass of its constituents. When protons and neutrons are assembled into a nucleus, mass is lost. The mass of a nucleus of carbon-12 is 1.992×10^{-27} kg, but the combined mass of its 6 protons and 6 neutrons is 2.008×10^{-27} kg. The carbon nucleus weighs less than the sum of its parts. The missing mass is known as the **mass difference**.

If we wished to pull a nucleus apart, back into its separate nucleons, the missing mass would have to be replaced. In fact we would need to put in energy in order to pull the nucleus apart. This energy is known as the **binding energy**, and it is also the energy released when a nucleus is formed from its constituent nucleons. Einstein's theory of special relativity is used to connect the mass defect, Δm, to the binding energy, E:

$$E = \Delta mc^2$$

where c is the speed of light $= 3 \times 10^8 \, \mathrm{m\,s^{-1}}$.

In the case of carbon-12 which has a mass difference of 1.6×10^{-29} kg, the binding energy is

$$E = 1.6 \times 10^{-29} \times 9 \times 10^{16} = 1.44 \times 10^{-12} \, \mathrm{J}$$

This is the amount of energy that would be needed to separate the 6 protons and 6 neutrons.

Large nuclei, like uranium-235, have a large mass difference and consequently have a large value of binding energy. However, this does not necessarily mean that they are stable. It is the **binding energy per nucleon** that tells us how much energy, on average, it will take to pull out each nucleon. The isotope with the highest value of binding energy per nucleon is iron-56, which is the most stable isotope.

Examiners' Notes

When you use the equation $E = \Delta mc^2$, Δm must be in kg to give E in joules.

Fig 16
Binding energy per nucleon vs nucleon number

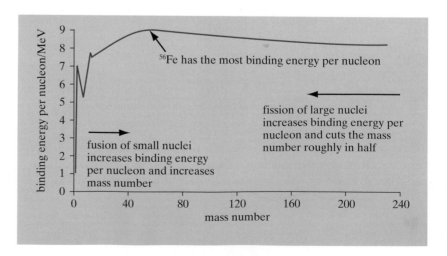

You can see from the graph in Fig 16 that very massive nuclei can increase their stability by splitting into two smaller isotopes. This is known as nuclear **fission** and certain isotopes of uranium and plutonium decay by spontaneously splitting into two smaller nuclei. A large amount of energy is released.

Energy is also released when two small nuclei merge to form a larger one. This is a nuclear **fusion**.

Energy from nuclear reactions

The energy released by nuclear reactions, such as alpha or beta decay, can also be explained by a change in mass. Radium-226 is an alpha particle emitter. It decays to radon, emitting an alpha particle with an energy of about 5 MeV:

$$_{88}^{226}\text{Ra} \rightarrow _{86}^{222}\text{Rn} + \alpha$$

The mass of the radium nucleus is $3.752\,02 \times 10^{-25}$ kg

The mass of the radon nucleus is $3.685\,49 \times 10^{-25}$ kg

The mass of the alpha particle is 0.06644×10^{-25} kg

When we compare the original mass of the radium nucleus with the mass of the reaction products, we find that there is a mass difference of 9.00×10^{-30} kg. It is this mass which has been transferred to energy. Using the equation $E = \Delta mc^2$, we can calculate the energy released by the equation:

$$E = 9.00 \times 10^{-30} \times 9 \times 10^{16} = 8.1 \times 10^{-13} \text{ J}$$

The usual SI units for mass and energy, kilograms and joules, are rather large for calculations on an atomic scale. We define smaller units, namely the **atomic mass unit, u**, and the **electron volt, eV**.

The atomic mass unit is defined as 1/12 of the mass of a carbon-12 atom. It is equal to 1.66×10^{-27} kg. Using this unit a hydrogen atom has a mass of approximately 1 u. Accurate values are given in Table 2 overleaf.

The electron volt is a more useful measure of energy on an atomic scale. 1 electron volt is equal to 1.60×10^{-19} J.

We can use Einstein's relation, $E = \Delta mc^2$, to calculate the energy equivalent of one atomic mass unit.

$$E = \Delta mc^2 = 1.66 \times 10^{-27} \times 9 \times 10^{16}$$
$$= 1.494 \times 10^{-10} \text{ J}$$
$$= \frac{1.494 \times 10^{-10}}{1.6 \times 10^{-19}} \text{ eV}$$
$$= 9.313 \times 10^8 \text{ eV or } 931.3 \text{ MeV}$$

This means that if 1 atomic mass unit of matter were entirely transferred to energy, it would release 931.3 MeV.

Essential Notes

Energy will be released by any nuclear reaction that results in products that have a higher binding energy per nucleon.

Essential Notes

Einstein's relation, $E = \Delta mc^2$, doesn't just apply to nuclear reactions. It links energy and mass changes for *all* physical systems. A moving object increases its mass as it accelerates. The relationship describes the energy needed to create particles in pair production, and the energy released by annihilation of matter and antimatter (see Unit 1).

Essential Notes

You previously used electron volts to calculate energy changes in atoms in Unit 1.

Examiners' Notes

A quick way of finding the energy released in MeV is to multiply the mass difference in atomic mass units, u, by the factor 931.3.

Table 2
Masses of some particles

Particle	Mass (kg)	Mass (u)	Energy equivalent (MeV)
Carbon-12 atom	1.99200×10^{-26}	12.0000	11175.6
Electron	9.11000×10^{-31}	5.48295×10^{-4}	0.51109
Proton	1.67208×10^{-27}	1.00728	938.080
Neutron	1.67438×10^{-27}	1.00867	939.374
Alpha particle	6.64250×10^{-27}	4.00151	3726.61

Nuclear fission

Nuclear fission occurs in a few heavy nuclei, which decay by splitting into two roughly equal fragments (known as the fission products) and a few free neutrons and beta particles. Spontaneous fission is rare but fission can be induced in uranium-235 by allowing it to absorb another neutron. The resulting isotope, uranium-236, is very unstable. It decays by fission, often emitting two or three extra neutrons at high speed. The energy released in such a reaction can be calculated by finding the mass difference. The equation for a typical fission is

$$^{235}_{92}\text{U} + ^{1}_{0}\text{n} \rightarrow ^{134}_{54}\text{Xe} + ^{100}_{38}\text{Sr} + 2^{1}_{0}\text{n}$$

The total mass was originally

$$235.044\,\text{u}\,(^{235}_{92}\text{U}) + 1.009\,\text{u}\,(^{1}_{0}\text{n}) = 236.053\,\text{u}$$

After the fission the mass is

$$133.9054\,\text{u}\,(\text{Xe}) + 99.9354\,\text{u}\,(\text{Sr}) + 2 \times 1.009\,\text{u}\,(2\text{n}) = 235.859\,\text{u}$$

The mass difference is 0.194 u, giving an energy release of 180 MeV per fission. This is a huge energy output compared with even the most energetic chemical reactions.

Nuclear fusion

The energy that powers stars comes from nuclear fusion, the joining of two light nuclei to make a more massive one. Fusion reactions also release energy. For example, consider the fusion of two deuterium nuclei to make helium:

$$^{2}_{1}\text{H} + ^{2}_{1}\text{H} \rightarrow ^{3}_{2}\text{He} + ^{1}_{0}\text{n}$$

The mass of two deuterium nuclei is $\quad 2 \times 2.01410 \quad = 4.02820\,\text{u}$

The mass of helium-3 + a neutron is $\quad 3.01605 + 1.00867 = 4.02472\,\text{u}$

This is a mass difference of 0.00348 u. We can convert this directly to energy in MeV:

$$\text{energy released} = 0.00348\,\text{u} \times 931.3\,\text{MeV} = 3.24\,\text{MeV released}$$

The energy released from fusion reactions is enormous, typically around 10^{14} joules per tonne of fuel. This is even larger than the energy released from fission reactions, and yet there are no fusion power stations.

Essential Notes

The fission products vary from fission to fission. A wide range of isotopes are produced. The energy released is typically 200 MeV per fission. Most of this is released instantly as kinetic energy of the fission products and neutrons, and energy of gamma rays. The remainder is released over a longer period of time as the unstable fission products decay, releasing beta-particles and neutrinos.

Essential Notes

Deuterium is an isotope of hydrogen.

Examiners' Notes

Calculations of mass difference involve subtracting two numbers to find the small difference between them. It is essential to retain a large number of significant figures until the end of the calculation.

There are complex technical problems to be overcome before fusion can be used as a commercially viable energy source. The two nuclei have to be brought very close, within 10^{-15} m, before they will fuse. At this distance the electrostatic repulsion between two positively charged protons is huge. The protons have to be moving very quickly if they are to overcome this repulsion. This requires very high temperatures, of the order of 100 million degrees Celsius. It has proved difficult to contain and control material at such high temperatures. All the nuclear power stations built so far have been fission reactors.

Essential Notes

At these very high temperatures all the electrons are stripped off their atoms. The material in an experimental fusion reactor is in the form of ionised gas, known as plasma.

Induced fission

Nuclear power stations produce about 15% of the electricity in the UK. This proportion is much higher in some other European countries. There are a number of different reactor designs but all the reactors use the energy released by nuclear fission, usually of uranium-235, to generate high-pressure steam. The steam drives turbines, which in turn drive electrical generators.

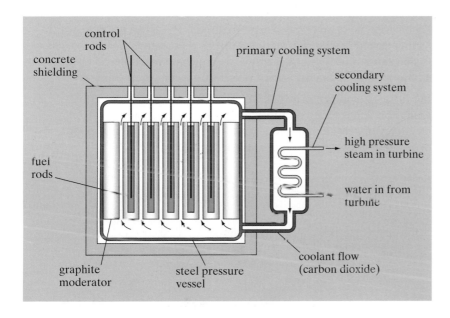

Fig 17
Gas-cooled nuclear reactor

Natural uranium is a mixture of several different isotopes. Uranium-238 is by far the most common (99.28%), whilst uranium-235 is much less common (0.718%). Uranium-238 is only slightly fissile when exposed to very high-energy neutrons, whereas uranium-235 readily undergoes fission with low-energy (thermal) neutrons. So it is the relatively rare uranium-235 that is the useful fuel for nuclear reactors. Some reactors use fuel in which the percentage of uranium-235 is increased. This is known as **enriched nuclear fuel**. Even so, the natural fission rate in uranium-235 is extremely slow. A practical nuclear reactor has to have a much greater power output. Bombarding the uranium-235 with neutrons can increase the fission rate. This is known as **induced fission**.

Fig 18
Pressurised water reactor

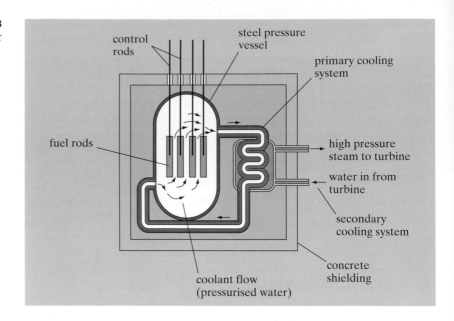

The chain reaction and critical mass

Each fission reaction releases extra neutrons. These can go on to induce further fission reactions. If all the neutrons that are released from each fission cause more fission reactions, then a chain reaction will ensue. A large amount of energy is then released in a very short time.

In practice not all the neutrons will cause further fission reactions. A neutron that is emitted from a fission reaction typically has an energy of 10 MeV, and is moving at a very high speed. A number of different things can happen to the neutron.

Fig 19
Chain reaction

One fission may release 3 neutrons. These can cause further fissions, releasing 9 neutrons, then 27 neutrons, and so on

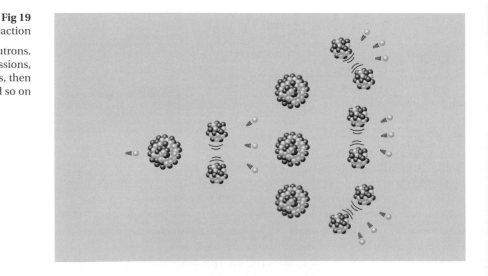

1. The neutron may leave the sample of uranium without causing any further reactions. In a small piece of uranium this is quite likely. In a larger piece of uranium, most of the neutrons will cause further fission. There is a **critical volume** for uranium below which a chain reaction cannot be sustained. For a spherical piece of uranium this corresponds to a **critical mass** of a few kilograms.

2. The neutron could be absorbed by uranium-238, or other nuclei, without causing any further fission.

3. The neutron could be absorbed by a nucleus of uranium-235 and cause another fission reaction.

Moderation and control

The probability that uranium-235 will absorb a neutron depends on the neutron speed. There is a much greater chance of absorption if the neutron is travelling at low speeds. It is important to slow the neutrons down quickly, so that they can cause more fission reactions. A material known as a **moderator** is used to slow the neutrons down.

In a nuclear reactor the fuel is in the form of hundreds of narrow rods surrounded by the moderator. These fuel rods each contain less than the critical mass of fissionable uranium. Most of the emitted neutrons leave the fuel rod and collide with the atoms of the moderator, slowing down in the process, before reaching the next fuel rod at the right speed to cause further fission reactions. The material used for a moderator has to have a low mass number so that as much as possible of the neutron's kinetic energy is transferred at each collision. Suitable materials are water, since it contains low-mass hydrogen atoms, and graphite (a form of carbon).

If the reactor is to transfer energy at a steady rate then, on average, each fission reaction must lead to one more fission reaction. If there are too

Essential Notes

In an atomic bomb, two or more pieces of uranium (or plutonium), each less than the critical mass, are pushed together violently by a conventional chemical explosion. Once together their combined mass is greater than the critical mass, a rapid chain reaction takes place and there is an enormous explosion.

Essential Notes

Slow neutrons are often referred to as thermal neutrons because their energy is equal to the average kinetic energy of atoms of the medium. At 20 °C this is about 0.025 eV.

Essential Notes

Imagine throwing a squash ball first at a football, then at a tennis ball. The football has a much larger mass and so will move off with a lower velocity. Since kinetic energy depends on velocity squared, the football will carry off less kinetic energy. The collision with the tennis ball will slow the squash ball down more. The most efficient transfer happens when both particles have the same mass.

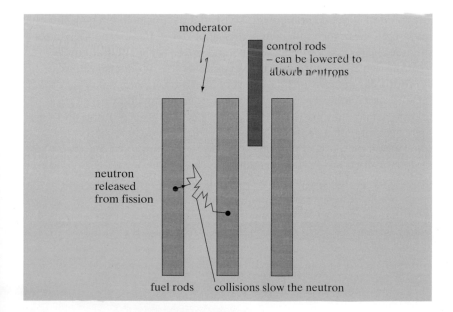

Fig 20
The roles of moderator and control rods in a nuclear reactor

many fission reactions occurring, **control rods** are used to absorb neutrons. Control rods are made of a material like boron or cadmium, which absorb neutrons well. These are lowered further into the reactor core to reduce the rate of fissioning, or raised to increase the rate.

Neutrons transfer energy to the moderator, which becomes very hot. This thermal energy is transferred to a coolant, often carbon dioxide or water, which in turn passes the energy on to a secondary coolant, water, that is turned to steam. The coolant has to be a fluid so that it can be pumped around the reactor core, and it should have a large specific heat capacity.

Safety aspects

It is impossible for a nuclear explosion to occur in a power station. The fuel is not enriched enough and it would be impossible for a critical mass of fissionable uranium to come together. However, there is a danger from other explosions, which could lead to an escape of radioactive matter into the environment.

The worst nuclear accident occurred at Chernobyl in the Ukraine in 1986 when the reactor power got out of control. The coolant boiled and blew the huge concrete lid off the power station. To prevent accidents like this occurring, power stations have a number of safety features.

There is often a set of control rods held out of the reactor on large electromagnets. If the temperature gets too high, perhaps because of a power failure to the coolant fans, the control rods drop in and shut down the reactor. Power stations also have the capability to flood the reactor with nitrogen gas or, in the last resort, water. This would absorb any spare neutrons and stop all fission occurring.

The high neutron flux from the reactor, and the high levels of radiation from the fuel and the fission products, mean that substantial shielding is necessary. Thick steel shields and several metres of concrete surround the reactor.

A fuel rod of enriched uranium is slightly radioactive before it goes into the reactor, but it is much more dangerous after it has spent some time in the reactor. The reaction products left behind by the fission of uranium are neutron-rich, highly unstable and very radioactive. Products such as strontium-90 and caesium-137 are very dangerous to humans. When the fuel rods are removed from the reactor they are dropped into a large pool of water where they are left to cool down. The high temperatures and the high levels of radioactivity are allowed to fall. Fuel rods are then transported for reprocessing where the unused uranium is recovered. The fission products are still very radioactive. These 'high-level' waste products are kept deep underground in geologically stable repositories.

Essential Notes

Specific heat capacity is the energy needed to heat 1 kg of a substance by 1 °C (see page 31).

3.5.3 Thermal Physics

Thermal energy

Specific heat capacity

When energy is transferred to an object as thermal energy, its internal energy increases. In general the increase in internal energy will cause an increase in temperature. The temperature change caused by a given amount of energy depends on the mass of the object and on the material it is made of. Some materials require a lot of energy to cause a small temperature rise, whereas other materials need less. This property of materials is known as the **specific heat capacity**, c; it is measured in joules per kilogram per degree kelvin, $J\,kg^{-1}\,K^{-1}$.

> **Definition**
>
> *The specific heat capacity of a material is the energy needed to cause a temperature rise of 1 K in a mass of 1 kg.*

The **heat capacity** of an object, a saucepan for example, is the energy required to raise its temperature by 1 K, in $J\,k^{-1}$.

'Specific' means the value per unit mass. In SI units, a unit mass is one kilogram, so specific heat capacity refers to 1 kg of a given material. Values for given materials can be looked up in data tables.

We can use the specific heat capacity to calculate the energy, ΔQ, required to heat any mass, m, of the material by any temperature rise, ΔT.

$$\Delta Q = mc\,\Delta T$$

> **Example**
>
> A 2.2 kW electric kettle is used to heat 1.5 litres (1.5 kg) of water. Assume that all the energy is transferred as heat in the water, ignore the energy needed to warm the kettle itself and any heat losses. How long will it take the kettle to bring the water, initially at 10 °C, to the boil?
>
> **Answer**
>
> The total energy required is:
> $\Delta Q = mc\,\Delta T = 1.5 \times 4190 \times (100 - 10) = 565\,650\,J$
>
> The electric kettle transfers 2200 joules every second.
>
> Time needed $= \dfrac{565\,650}{2200} = 260\,s$ (to 2 s.f.)

The specific heat capacity of water is relatively high. This means that water is useful for transferring energy, for example in water-cooled engines and in nuclear power stations. The high specific heat capacity of water also means that the temperature of a large mass of water, like the sea, only changes slowly.

Essential Notes

A temperature change of 1 degree kelvin, 1 K, is identical to a temperature change of 1 °C. The Kelvin and Celsius scales have equal increments, they just start at different points.

$0\,K = -273.15\,°C$

$0\,°C = 273.15\,K$

See 'Temperature scales', page 35.

Material	Specific heat capacity ($J\,kg^{-1}K^{-1}$)
air	993
water	4190
copper	385
concrete	3350
gold	135
hydrogen	14 300

Table 3
Examples of specific heat capacities

Essential Notes

The English scientist James Joule demonstrated that kinetic energy could be transferred to thermal energy by using horses to turn paddle wheels in a tank of water. A temperature rise of the water showed that energy had been transferred.

Energy transfers

When energy is transferred from one form to another, for example from gravitational potential energy to kinetic energy, some of the energy inevitably ends up as random thermal energy in the surroundings. When a hammer is used to strike metal, potential energy is transferred to kinetic energy, but the end result is that the hammer and the metal get hotter. The brakes in a car transfer kinetic energy to thermal energy and the brakes can get red hot in the process.

Example

A laboratory method of demonstrating this energy transfer uses lead shot in a closed cardboard tube. The tube is held vertically and turned end to end so that the lead is continually being lifted and allowed to fall, hitting the bottom of the tube. The tube is 1 m long and there is 250 g of shot in it. Lead has a specific heat capacity of 126 J kg^{-1}K^{-1}. How much of a temperature rise would you expect in the lead after 100 inversions of the tube?

Answer

Each time the lead falls, potential energy is transferred to kinetic energy and then to thermal energy. The change in potential energy = $mg\Delta h$ = 0.250 kg × 9.81 × 1 m = 2.45 J, so 245 J after 100 inversions.

If this is all transferred as thermal energy in the lead, the temperature rise, ΔT, will be

$$\Delta T = \frac{\Delta Q}{mc} = \frac{245\,\text{J}}{(0.250\,\text{kg} \times 126\,\text{J kg}^{-1}\,\text{K}^{-1})} = 7.8\,^{\circ}\text{C}$$

In practice the temperature rise would be less than this because of heat losses to the surroundings.

Specific latent heat

When energy is transferred to an object as heat it does not always lead to an increase in temperature. The energy transfer can lead to a change in state, such as when ice melts or water turns to steam. The energy needed to change the state of a substance is known as its **latent heat, l**. The energy required to melt 1 kg of a solid is known as the **specific latent heat of fusion**. The energy required to convert 1 kg of a liquid to a gas is known as the **specific latent heat of vaporisation**.

Definition

The specific latent heat of fusion of a substance is the energy required to change 1 kg of a solid into 1 kg of liquid, with no change in temperature.

Definition

The specific latent heat of vaporisation of a substance is the energy required to change 1 kg of a liquid into 1 kg of gas, with no change in temperature.

The energy, ΔQ, needed to change the state of a substance of mass m, is therefore:

$$\Delta Q = ml$$

Material	Vaporisation (kJ kg^{-1})	Fusion (kJ kg^{-1})
water	2260	334
oxygen	243	14
helium	25	5
mercury	290	11
iron	6339	276
lead	854	25

Table 4
Specific latent heat values

When a substance changes from a solid to a liquid, or from a liquid to a gas, energy is needed to do work against the attractive forces holding the solid or liquid together. However, during a change of state there is no increase in the average kinetic energy of the particles, and so there is no change in temperature. Usually there is an increase in volume as a solid melts, or a liquid boils. Energy is needed to do work against external pressure as the substance expands. The latent heat of a substance is therefore the sum of the energy needed to increase the potential energy of its particles and to do work against external pressure.

Essential Notes

When water turns to steam at atmospheric pressure (1×10^5 Pa) about 7% of the energy supplied is needed to do work against atmospheric pressure; the rest is needed to increase the potential energy of its molecules as they move apart from each other.

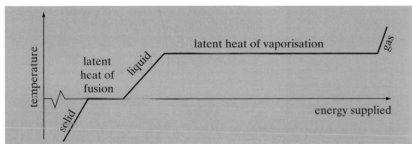

Fig 21
The temperature change of a substance as it is heated

If energy is supplied at a constant rate to a solid substance its temperature will rise until it reaches its melting point. The substance will then melt at constant temperature. When all the solid has turned into liquid, the temperature will rise again until the liquid reaches its boiling point. The substance will then boil at constant temperature until all the liquid has turned into gas, when the temperature will rise again.

The energy needed to change the state of a substance is often used to dissipate thermal energy. The cooling towers of a power station transfer energy from the power station by evaporating large amounts of water. We do the same thing on a smaller scale – when we sweat we are transferring excess thermal energy by evaporating liquid.

Example

An ice cube of mass 20 g is added to 200 g of water which is initially at 20 °C. As the ice melts it has a cooling effect on the water. If all the energy needed to melt the ice comes from the water, calculate the final temperature of the water.

Answer

The latent heat necessary to melt the ice is

$$\Delta Q = ml = 0.020 \text{ kg} \times 334 \times 10^3 \text{ J kg}^{-1} = 6680 \text{ J}$$

If this energy all comes from the water:

$$\frac{\Delta Q}{mc} = \Delta T = \frac{6680 \text{ J}}{(0.200 \text{ kg} \times 4190 \text{ J kg}^{-1} \text{ K}^{-1})}$$

$$\Delta T = 8.0 \text{ °C}$$

The temperature of the water will drop to around 12 °C because of the ice melting.

Note however that we have neglected here the thermal energy needed to raise the temperature of the 20 g of water from the melted ice at 0 °C to the final temperature of the water. If this also comes only from the water, the final temperature will be about 11 °C.

Example

The amount of water that we lose through the evaporation of sweat depends on the temperature of our surroundings, as well as on the humidity and wind speed. On average a typical person loses about 0.5 litre of sweat per day. Calculate the average power of this energy transfer.

Answer

The energy needed to evaporate 0.5 litre of sweat is $\Delta Q = ml$.

1 litre of water has a mass of 1 kg, so $m = 0.5$ kg.

l is the specific latent heat of vaporisation of water = $2.260 \times 10^6 \text{ J kg}^{-1}$.

So the energy transferred, $\Delta Q = 0.5 \text{ kg} \times 2.260 \times 10^6 \text{ J kg}^{-1} = 1.13 \times 10^6 \text{ J}$.

If this is transferred over a period of 24 hours or $24 \times 60 \times 60 = 86\,400$ seconds, the average power is:

$$\frac{1.13 \times 10^6}{86\,400} = 13 \text{ W}$$

Ideal gases

The physical state of a fixed mass of gas can be described by three physical quantities: pressure, temperature and volume.

Pressure The pressure, p, that a gas exerts on the walls of its container is caused by the collisions of the molecules with the walls. Pressure is defined as the force per unit area and is measured in pascals, Pa. 1 Pa is a pressure of 1 newton per square metre, $1\,Pa = 1\,N\,m^{-2}$.

Temperature   The temperature, T, of the gas is a measure of the average kinetic energy of its molecules. T is the absolute temperature measured in degrees kelvin, K.

Volume   The volume, V, is the space occupied by the gas, and is measured in m^3.

Temperature scales

The temperature of a gas is measured by the change in the property of some other substance at the same temperature. This is known as a **thermometric property**. For example, a liquid in glass thermometer measures temperature in terms of the expansion of a length of liquid, usually mercury or alcohol.

The thermometric property, length in the case of mercury thermometer, is measured at two **fixed points**. The change of state of a substance is often used as a fixed point in defining a temperature scale since the temperature is constant during a change of state. The Celsius temperature scale uses the change of state of pure water as its fixed points. 0 °C is defined as the melting point of pure ice and 100 °C is defined as the boiling point of pure water at a pressure of 1 atmosphere. The thermometric property, such as the length of a mercury thread, is assumed to vary linearly between these points and the scale is divided into 100 equal divisions.

Essential Notes

The resistance of a metal wire, the wavelength of radiation emitted by a body and the volume of a gas are all properties which are used to measure temperature.

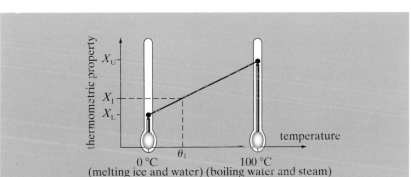

Fig 22
Defining a temperature scale

Unfortunately two different thermometers, perhaps one using mercury and one using alcohol, are only guaranteed to agree at the fixed points. This is because each thermometric property has its own temperature dependence; alcohol and mercury expand at different rates as they warm up. To overcome the problem of different temperature scales, a standard scale has been defined. This is the **absolute temperature scale**, also known as the **Kelvin scale**. The Kelvin scale is based on the behaviour of an **ideal gas**, which is a gas where there are no forces between its molecules (see page 41).

As an ideal gas is cooled, its molecules slow down and the gas exerts a lower pressure on the walls of its container. If the container has a fixed volume, the pressure of the gas will continue to drop as the temperature drops.

Essential Notes

The forces between the molecules in a gas are relatively small because the molecules are much further apart than those of a liquid or a solid. A dry gas at low pressure behaves as an ideal gas.

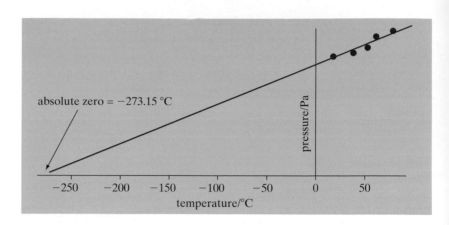

Fig 23
Pressure vs temperature for an ideal gas

Essential Notes

It is theoretically impossible to cool something to absolute zero, but researchers have got very close. Liquid helium has been cooled to 90 µK (1 µK is 1 microkelvin or 10^{-6} K).

Essential Notes

The **triple point of water** is the unique combination of pressure and temperature when ice, water and water vapour all exist in thermal equilibrium. This only happens when the pressure is 0.6% of atmospheric pressure and the temperature is 0.01 °C.

Essential Notes

At low temperatures, real gases can liquefy under pressure.

A graph plotted of pressure against temperature can be extrapolated backwards to very low temperatures. Eventually the pressure of an ideal gas would drop to zero. At this point the molecules have stopped moving and the gas cannot get any colder. This point is the lowest conceivable temperature and it is known as **absolute zero**.

The Kelvin scale uses the pressure of an ideal gas as its thermometric property. The two fixed points of the Kelvin scale are absolute zero and the **triple point of water**, defined to be 273.16 degrees on the Kelvin scale. This odd number is chosen to make a temperature change of one degree Celsius the same as one kelvin. On the Kelvin scale absolute zero is 0 K (-273.15 °C) and the ice-point is 273.15 K (0 °C).

> **Definition**
>
> *Temperature in kelvin = temperature in degrees Celsius + 273.15*

The gas laws

Most gases behave in similar ways. When the gas molecules are a long way apart, that is when the gas is at high temperature and low pressure, all gases behave as ideal and they are found to obey certain laws.

1. Boyle's law

When a gas is put under pressure, its volume decreases. For a fixed mass of an ideal gas at constant temperature, Boyle's law states that its volume is inversely proportional to its pressure.

> **Definition**
>
> *For a fixed mass of an ideal gas at constant temperature* $p \propto \dfrac{1}{V}$

For a fixed mass of gas at constant temperature therefore

$$pV = \text{constant} \quad \text{or} \quad p_1V_1 = p_2V_2$$

Example

$20 \, cm^3$ of air at atmospheric pressure, $1 \times 10^5 \, Pa$, is trapped in a bicycle pump when a finger is placed over the end of the pump. If the piston is pushed in until the volume is $5 \, cm^3$, find the new pressure.

Answer

$p_1 V_1 = p_2 V_2$

$20 \, cm^3 \times 1 \times 10^5 \, Pa = 5 \, cm^3 \times p_2$

$p_2 = 4 \times 10^5 \, Pa$

Examiners' Notes

Boyle's law only applies to a fixed mass of gas at constant temperature. You couldn't apply $P_1 V_1 = P_2 V_2$ to a tyre or a balloon that was being inflated since the new pressure and volume are affected by the increased number of gas molecules.

Fig 24
p vs V for an ideal gas at constant I

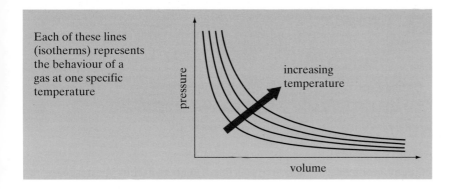

Each of these lines (isotherms) represents the behaviour of a gas at one specific temperature

increasing temperature

pressure

volume

2. Charles' law

When gases are heated at constant pressure, they all expand at the same rate. The volume of a gas is proportional to its temperature, providing we use the Kelvin temperature scale.

Definitions

The volume of a fixed mass of an ideal gas at constant pressure is proportional to its absolute temperature, V ∝ T.

3. The pressure–temperature law

If a gas is heated in a container of fixed volume its pressure will increase. In fact the pressure of a gas is proportional to its absolute temperature.

Definitions

The pressure of a fixed mass of an ideal gas at constant volume is proportional to its absolute temperature, P ∝ T.

The equation of state

For a fixed mass of gas, the three gas laws can be summarised as:

$p \propto T$ at constant volume

$V \propto T$ at constant pressure

$p \propto \dfrac{1}{V}$ at constant temperature

It is possible to combine these three laws into an equation of state for an ideal gas:

$$pV \propto T \quad \text{or, for 1 mole of ideal gas,} \quad pV = RT$$

The constant R is known as the **molar gas constant**: $R = 8.31\,\text{J}\,\text{mol}^{-1}\,\text{K}^{-1}$. For n moles of gas the equation becomes:

$$pV = nRT$$

Essential Notes

A **mole** is the SI unit of amount of substance. One mole of a gas contains 6.02×10^{23} molecules.

Definition

The **equation of state for an ideal gas**, or **ideal gas equation**, is $pV = nRT$.

In many problems the mass of gas is fixed, such as in the cylinder of an engine or in a sealed balloon or car tyre. The initial values of the gas's pressure, volume temperature can be written, p_1, V_1 and T_1 and the equation of state can be written:

$$\frac{p_1 V_1}{T_1} = nR$$

If the gas is compressed or heated then its pressure, volume and temperature will change to new values, p_2, V_2 and T_2, but n remains fixed. The equation of state is now:

$$\frac{p_2 V_2}{T_2} = nR$$

This gives a useful form of the equation:

$$\frac{p_1 V_1}{T_1} = \frac{p_2 V_2}{T_2}$$

Example

A bubble of air escapes from a diver's breathing apparatus at a depth of 45 m. The bubble has a volume of $2.0 \times 10^{-5}\,\text{m}^3$. The water pressure at a depth of 45 m is 450 kPa and the water temperature is 5 °C. What is the volume of the bubble when it has risen to the surface, where the temperature is 10 °C? Take atmospheric pressure as 100 kPa.

Answer

The total pressure at a depth of 45 m is 550 kPa, 450 kPa due to the water and 100 kPa due to atmospheric pressure.

The initial conditions of the gas are: $p_1 = 550$ kPa; $V_1 = 2.0 \times 10^{-5}$ m^3; $T_1 = 278$ K

The final conditions of the gas are $p_2 = 100$ kPa; V_2 is unknown and $T_2 = 283$ K

The mass of air in the bubble is fixed so

$$\frac{p_1 V_1}{T_1} = \frac{p_2 V_2}{T_2}$$

$$V_2 = \frac{p_1 V_1 T_2}{p_2 T_1} = \frac{550 \text{kPa} \times 2.0 \times 10^{-5} \text{m}^3 \times 283 \text{K}}{100 \text{kPa} \times 278 \text{K}}$$

$$V_2 = 1.1 \times 10^{-4} \text{m}^3$$

The Avogadro constant, the molar gas constant and the Boltzmann constant

Avogadro's law states that, at the same temperature and pressure, equal volumes of gases contain equal numbers of molecules. This is in agreement with the equation of state for an ideal gas. Since $pV = nRT$, then

$$n = \frac{pV}{RT}$$

If the pressure, volume and temperature are the same for any two gases then n, the number of moles, must also be the same. Since a mole of substance always contains the same number of particles, Avogadro's law and the ideal gas equation are equivalent alternatives.

One mole of a substance contains as many particles (these could be atoms, ions, electrons or molecules) as there are atoms in 12 g of the carbon isotope $^{12}_{6}$C. The number of molecules in a mole of gas is always 6.02×10^{23}.

This number is known as the **Avogadro constant**, and is written N_A.

Definition

The Avogadro constant, N_A, is the number of particles in a mole of substance.

$N_A = 6.02 \times 10^{23}$

The molar mass, that is the mass of 1 mole of a substance, is its relative molecular mass expressed in grams. A mole of hydrogen therefore has a mass of only 1 g whilst a mole of carbon-12 has a mass of 12 g.

Note that the molecular mass is linked to the molar mass by the Avogadro constant:

$$\text{molar mass} = \text{molecular mass} \times \text{Avogadro constant}$$

so that the molecular mass of carbon-12 is:

$$\frac{\text{molar mass}}{N_A} = \frac{12}{6.02 \times 10^{23}} = 1.99 \times 10^{-23} \text{ g or } 1.99 \times 10^{-26} \text{ kg}$$

We can relate the pressure, volume and temperature of a gas to the *number of molecules* in the sample of gas, N, rather than the number of moles, n. Since $pV = nRT$, and $n = N/N_A$, the ideal gas equation can be written:

$$pV = \frac{NRT}{N_A}$$

The **Boltzmann constant**, k, is defined as the molar gas constant divided by the Avogadro constant.

Definition

The Boltzmann constant, k, is the molar gas constant, R, divided by the Avogadro constant, N_A.

$$k = \frac{R}{N_A}$$

k has a value of $1.38 \times 10^{-23} \text{J K}^{-1}$.

So, for N molecules of gas, the ideal gas equation can be written:

$$pV = NkT$$

Example

A car tyre has a volume of around $1.50 \times 10^{-2} \text{ m}^3$ and it has been inflated to a pressure that is twice atmospheric pressure. If the temperature is $20\,°C$, estimate the mass of air in the tyre. (The relative molecular mass of air is about 29; atmospheric pressure is $100\,\text{kPa}$.)

Answer

$$\frac{pV}{RT} = n = \frac{200 \times 10^3 \text{Pa} \times 1.50 \times 10^{-2} \text{m}^3}{8.31 \text{ J mol}^{-1} \text{K}^{-1} \times 293 \text{ K}} = 1.23 \text{ mole}$$

A mole of air would have a mass of 29 g so the mass of air in the tyre is about $1.23 \times 29 = 36\,\text{g}$.

Molecular kinetic theory model

The average kinetic energy of molecules in a gas

As the molecules of a gas move and collide, the total kinetic energy is shared amongst the molecules. The velocity of any particular molecule in a gas is random, it may be in any direction and it may have any one of a range of values. However the average kinetic energy of the molecules is proportional to the temperature of the gas. Kinetic energy is calculated as $\frac{1}{2}mv^2$, so if the average (mean) squared speed of a gas molecule is written $\overline{c^2}$ (the bar denotes the mean value and the expression is pronounced 'c-squared bar') we can write:

$$T \propto \frac{1}{2}m\overline{c^2}$$

This suggests that at absolute zero the molecules have no kinetic energy. In other words they have stopped moving. In fact this is never quite the case, since quantum effects lead to a zero-point energy, which means that the kinetic energy never falls entirely to zero.

Pressure of an ideal gas

An ideal gas is one which obeys the equation of state, $pV = nRT$, exactly. There are no gases which are perfectly 'ideal', but real gases at low pressures and at temperatures well above their liquefying temperature come quite close to ideal behaviour. It is possible to imagine a theoretical gas that would obey the gas laws exactly. This theoretical gas has to have the following properties:

- Its molecules are negligibly small. We can ignore the space taken up by the molecules in comparison to the space between molecules. In other words the total volume occupied by the molecules is very much smaller than the volume occupied by the gas.

- Its molecules have elastic collisions. There is no loss of kinetic energy when a gas molecule collides with the walls of its container.

- The collisions take very little time, compared to the time between collisions.

- The molecules do not exert any force on each other, except during collisions. There are no intermolecular forces of attraction.

- There are a large number of molecules, so that a statistical approach can be used.

- The molecules move in random directions.

These assumptions are used to develop a **kinetic theory** of gases, which tries to predict the behaviour of a gas by considering the motion of its molecules. In particular we can derive an expression which links the pressure exerted by a gas with the speed of its molecules.

We start by imagining a single molecule, alone in a cubical box with sides of length l metres (Fig 25). The molecule has a mass m and is moving in a random direction with velocity c. The molecule will keep moving in a straight line, obeying Newton's first law of motion, until it has a collision

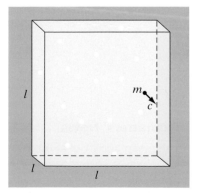

Fig 25

41

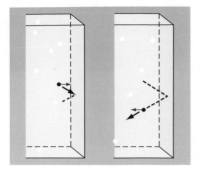

Fig 26

with one of the walls of the container. It will exert a force on the wall before it bounces back and travels in the opposite direction (Fig 26). If we can find the force, and hence the pressure, exerted by one molecule, we can add up all the forces exerted by all the molecules in the box.

To simplify things a little further we will consider only the component of the velocity in the x-direction, that is perpendicular to the wall. The force exerted by a single collision can be found from Newton's second law of motion (see Unit 2). The molecule has an initial momentum, mv_x, and after an elastic collision with wall the momentum will be the same size, but in the opposite direction, $-mv_x$.

The change in momentum = final momentum − initial momentum

$$= -mv_x - mv_x = -2mv_x$$

The molecule will hit this wall again after it has travelled to the other side of the box and back again, a distance of $2l$.

This will take a time of $\dfrac{2l}{v_x}$.

Force is defined by Newton's second law as the rate of change of momentum, so the force, F, exerted on the molecule is:

$$F = \frac{\text{change in momentum}}{\text{time taken}} = \frac{-2mv_x}{2l/v_x} = -\frac{mv_x^2}{l}$$

By Newton's third law, the force exerted *on the wall* is equal to this but in the opposite direction. Pressure is force per unit area. The area of one wall is l^2, so

$$p = \frac{mv_x^2}{l^3}$$

Since l^3 is the volume, V, of the box, this equation becomes:

$$p = \frac{mv_x^2}{V}$$

The total pressure is the pressure from all of the other particles as well:

$$p = p_1 + p_2 + p_3 + \ldots \quad = \frac{mv_{x1}^2}{V} + \frac{mv_{x2}^2}{V} + \frac{mv_{x3}^2}{V} + \ldots$$

$$= \frac{m(v_{x1}^2 + v_{x2}^2 + v_{x3}^2 + \ldots)}{V}$$

The sum in the brackets is the total of all the molecules' squared velocities. This is equal to the mean squared speed, $\overline{v_x^2}$, multiplied by the total number of molecules, N, in the box. So the equation becomes:

$$p = \frac{Nm\overline{v_x^2}}{V}$$

Examiners' Notes

The mean *velocity* of all the molecules is always zero, because they move randomly in all directions.

This is the pressure on a wall due to the x-component of the velocity of all the molecules in the box. Finally we must take into account the other components of velocity.

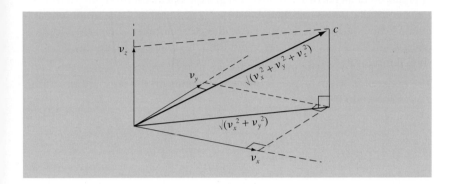

Fig 27

The speed of the particle c is linked to its components in the x, y and z directions by Pythagoras' theorem (see Fig 27):

$$c^2 = v_x^2 + v_y^2 + v_z^2$$

Because the motion is random, the molecules are equally likely to be moving in any of the three directions so the mean value of v_x^2, v_y^2 and v_z^2 will be the same: $\overline{c^2} = 3\overline{v_x^2}$.

So we now have this expression for the pressure:

$$p = \frac{Nm\overline{c^2}}{3V} \quad \text{or} \quad pV = \frac{1}{3}Nm\overline{c^2}$$

This equation links the pressure and volume of a gas to the number of molecules and their mean squared speed.

If we define a quantity c_{rms} as the **root-mean-square (r.m.s.) speed** of the molecules, that is

$$c_{rms} = \sqrt{\frac{(c_1^2 + c_2^2 + c_3^2 + ...)}{N}}$$

then

$$c_{rms^2} = \frac{(c_1^2 + c_2^2 + c_3^2 + ...)}{N} = \overline{c^2}$$

and we can write the above equation as

$$pV = \frac{1}{3}Nmc_{rms^2}$$

Since Nm is the total mass of the gas, Nm/V is actually the density, ρ, of the gas. The equation can be written:

$$p = \frac{1}{3}\rho\overline{c^2} = \frac{1}{3}\rho c_{rms^2}$$

Internal energy: relation between temperature and average molecular kinetic energy

The **internal energy** of a substance is the sum of the potential and kinetic energies of all its particles. In an ideal gas the particles are so far apart that intermolecular forces can be disregarded. The particles have no potential energy. The internal energy of a gas is therefore entirely due to the kinetic energy of the molecules. At any given time the total kinetic energy is shared randomly between all the molecules in the gas.

The range of molecular speeds on a gas depends on the temperature of the gas, as shown in Fig 28.

Fig 28
The range of molecular speeds at a given temperature is given by the Maxwell-Boltzmann distribution

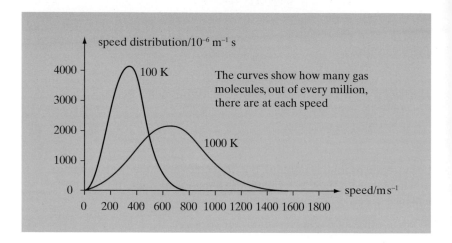

The curves show how many gas molecules, out of every million, there are at each speed

Examiners' Notes

You need to know that, although the actual speed of a given molecule is random within the range, the most probable speed increases at higher temperatures.

The kinetic theory of gases gives a microscopic view of what is happening within a gas. It allows us to link the temperature of a gas with the average kinetic energy of the gas molecules.

For 1 mole of gas, kinetic theory leads to the equation:

$$pV = \frac{1}{3} N_A m c_{rms}^2$$

The equation of state for an ideal gas gives a macroscopic view; it links the pressure and volume of a gas to its temperature. For 1 mole of gas:

$$pV = RT$$

These two expressions are equivalent and so:

$$\frac{1}{3} N_A m c_{rms}^2 = RT$$

This equation then lets us link the temperature of a gas to the average kinetic energy of its molecules, which is $\frac{1}{2} m c_{rms}^2$:

$$\frac{1}{2} m c_{rms}^2 = \frac{3}{2}\left(\frac{R}{N_A} \right) T$$

The ratio R/N_A is the Boltzmann constant, k (see page 40.)

So the equation can be written:

$$\frac{1}{2}mc_{rms}^2 = \frac{3}{2}kT$$

Example

Calculate the average kinetic energy of the molecules of a gas at 20 °C. If the gas is oxygen, find the r.m.s. speed of the molecules. The relative molecular mass of oxygen is 32.

Answer

$$\frac{1}{2}mc_{rms}^2 = \frac{3}{2}kT = 1.5 \times 1.38 \times 10^{-23}\,\text{J K}^{-1} \times 293\,\text{K} = 6.07 \times 10^{-21}\,\text{J}$$

If the mass of 1 mole of oxygen is 32×10^{-3} kg, then the mass of 1 molecule =

$$\frac{32 \times 10^{-3}}{N_A} = 5.32 \times 10^{-26}\,\text{kg}$$

This gives a value for the mean squared speed of:

$$C_{rms}^2 = \frac{2 \times 6.07 \times 10^{-21}\text{J}}{5.32 \times 10^{-26}\,\text{kg}} = 2.28 \times 10^5\,\text{m}^2\text{s}^{-2}$$

Therefore the r.m.s. speed of an oxygen molecule at 20 °C is about $480\,\text{m s}^{-1}$.

How Science Works

As well as understanding the physics in this unit, you are expected to develop an appreciation of the nature of science, the way that scientific progress is made and the implications that science has for society in general. GCSE and A-level science syllabuses refer to these areas as '*How Science Works*'.

The *How Science Works* element of your course, which also includes important ideas about experimental physics, may be assessed in the written examination papers as well as in the internally assessed unit, the Investigative Skills Assessment or ISA. The concepts included in *How Science Works* may be divided into several areas.

Theories and models

Physicists use theories and models to attempt to explain their observations of the universe around us. These theories or models of the real world can then be tested against experimental results. Scientific progress is made when experimental evidence is found that supports a new theory or model.

You are expected to be aware of historical examples of how scientific theories and models have developed and how this has changed our knowledge and understanding of the physical world.

An example of a scientific model is the kinetic theory of gases. An ideal gas is modelled by making several simplifying assumptions, such as that the molecules of the gas occupy zero volume. This is close to the truth, since the total volume occupied by the molecules is very much less than the volume of the gas itself. We also assume that there are no forces between the molecules except during collisions. Again this is almost true since the intermolecular forces are short range and the gas molecules are, on average, a relatively long way apart. When we apply Newton's laws of motion to the molecules of this ideal gas we are able to derive expressions for temperature and pressure which are in close agreement with the experimentally derived gas laws, such as Boyle's law.

Experimental results can of course disprove a model that was previously accepted. Rutherford's scattering experiment showed that the 'plum pudding' model of an atom could not be correct. Rutherford's results led to the development of a new model, that of the nuclear atom.

You should know the meaning of the terms 'hypothesis' and 'prediction'.

- A **hypothesis** is a tentative idea or theory, or explanation of an observation.

- A **prediction** from a hypothesis or theory is a forecast that can be tested by experiment.

If a reliable experiment does not support a hypothesis, then the hypothesis is likely to be abandoned or modified. Hypotheses are not usually widely accepted until the experimental results have been repeated by a number of independent scientists. It may take many experimental tests until a set of hypotheses become accepted as a scientific theory (see the figure left).

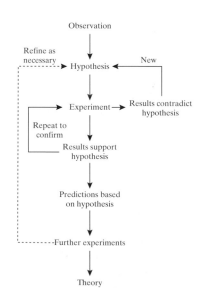

The stages of scientific research

Even then, a scientific theory is always capable of being later refuted if compelling experimental evidence suggests that a new explanation is necessary.

Experimental techniques

You are expected to develop the skills of experimental planning, observation, recording and analysis. These skills will mainly be assessed in the practical coursework, the ISA, or by the externally-marked practical assignment, the EMPA. This section contains some general advice for carrying out experimental work in physics.

When you plan an experiment you need to be able to identify the dependent, independent, and control variables that are involved.

- The **independent variable** is the physical quantity that you deliberately change.

- The **dependent variable** changes as a result of this.

For example, if you are asked to investigate how the length of a piece of metal wire affects its electrical resistance, the length is the independent variable, and the electrical resistance is the dependent variable. Any other variables that may have an impact on the outcome need to be controlled so that the conclusions of the experiment are clear. These are known as the **control variables**. In the example of the wire, two of the control variables are the cross-sectional area of the wire and its temperature.

You will need to select appropriate apparatus, including measuring instruments of a suitable precision and accuracy. These two terms are often confused.

- **Accuracy** refers to how close the reading is to the accepted value.

- **Precision** refers to the number of significant figures that the measurement is made to.

For example, an electronic balance that gives an answer to 0.01 g, e.g. 3.24 g, is capable of more precise measurements than a balance that measures in grams only, e.g. 3 g. If several readings of the same measurement are closely grouped together, the readings are said to be **precise**. If the readings agree with a known mass, they are said to be **accurate**. The analogy of rifle shots at a target may be used to differentiate between accuracy and precision (see the figure right).

A accurate

Figure A shows accurate shooting, since the bullets (or readings) are close to the centre of the target. But the shooting is not precise, since the bullets are widely scattered. Figure B shows precise shooting, since the bullets (readings) are closely grouped, but not accurate, since the bullets (readings) are not close to the centre of the target (accepted value).

B precise

When choosing your apparatus you should be aware that ICT can be used to assist with the collection and analysis of experimental data. This may mean using a suitable sensor, attached to a data logger, to take the readings and then a spreadsheet to help to analyse them. For example, suppose that you wanted to investigate the current surge that passes through the filament of an incandescent light bulb when it is first turned on. The light

Using logarithms and logarithmic graphs

In your A2 examinations, you may be asked to construct or interpret graphs with a logarithmic scale. If you are studying A-level Mathematics you may have met logarithms already. This section tells you what you need to know for A-level Physics.

There are two main reasons for using logarithmic scales on graphs in physics. First, a graph with logarithmic scales enables you to show a much wider range of values, e.g. from 10 to 10 000, which would be impossible to show on a linear scale. Second, a logarithmic graph enables you to find an unknown power in an equation linking two variables, such as n in $y = x^n$.

The logarithm of a number to the base 10 is the power to which 10 must be raised to equal that number. Suppose that $x = 10^y$, we say that y is the logarithm of x to the base 10:

If $x = 10^y$ then $y = \log_{10}x$

For example, $100 = 10^2$ so the logarithm of 100 (to the base 10) is 2.

Suppose you had to plot a graph of the data shown in Table 1.

If you used an x-scale that allowed you to plot the first through to the last value on the x-axis (with 1 mm representing 1 amp), you would need rather large graph paper (3 m long!). It would need to be even larger to fit all the y values. Using logarithmic scales on both axes solves this problem.

Using your calculator to find the logarithms of the current and power values, you can add two new columns to the table – as shown in Table 2.

Table 1

Current/A	Power/W
1	1
10	100
100	10 000
1000	1 000 000
2000	4 000 000
3000	9 000 000

Table 2
Note the headings on the new columns – a logarithm is an index (power) and it has no units, even if the original quantity did have units. 'Log' is used to indicate 'logarithm to the base 10'.

Current/A	Power/W	log(current/A)	log(power/W)
1	1	0.00	0.00
10	100	1.00	2.00
100	10 000	2.00	4.00
1000	1 000 000	3.00	6.00
2000	4 000 000	3.30	6.60
3000	9 000 000	3.48	6.95

This data can now be plotted as a log–log graph.

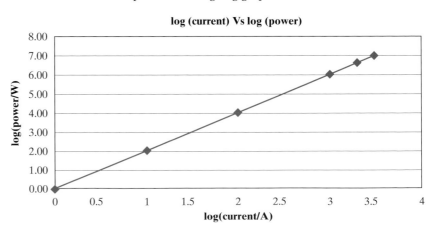

log (current) Vs log (power)

Logarithms transform multiplication to the process of addition:

$$\log_{10}(AB) = \log_{10}A + \log_{10}B \qquad \text{Equation 1}$$

This is because when we multiply two powers together we add the indices, e.g. $10^2 \times 10^3 = 10^5$.

Another feature of logarithms is that the process of raising a number to a certain power is transformed to multiplication:

$$\log_{10}(A^n) = n \times \log_{10}A \qquad \text{Equation 2}$$

This is because $(10^2)^3 = 10^2 \times 10^2 \times 10^2 = 10^6$, or $10^{2 \times 3}$.

These two properties of logarithms can be used to plot graphs to find unknown powers. For example, it is known that the force, F, between two bar magnets varies with the distance between them, x, according to the equation:

$$F = Ax^n$$

where A and n are unknown constants.

A graph of F against x would produce an exponential curve, so it is useful to use logarithms. Taking logs of both sides:

$$\log F = \log(Ax^n)$$

Using equation 1, $\log F = \log A + \log x^n$.

Using equation 2, $\log F = \log A + n \log x$.

Compare this to the equation of a straight line graph:

$$y = mx + c$$

If we plot $\log F$ on the y-axis and $\log x$ on the x-axis, the gradient (m) will give the power, n, and the intercept will be equal to $\log A$.

In Physics A-level studies you will often need to use logarithms to the base 10. However, it is sometimes useful to take logarithms to the base e, where e = 2.718. These are called natural logarithms. They are written in the form $\log_e x$, or $\ln x$ ('ln' means natural logarithm). The term $\ln x$ is the inverse function of e^x, so that $\ln e^x = x$.

Natural logarithms are used to plot exponential relationships, such as capacitor discharge and radioactive decay. For example, to find the time constant, CR, from capacitor discharge data giving the potential difference, V, at different times, t:

$$V = V_0 e^{\frac{-t}{CR}}$$

$$\frac{V}{V_0} = e^{\frac{-t}{CR}}$$

Taking natural logs of both sides gives:

$$\ln\frac{V}{V_0} = \frac{-t}{CR}$$

so that CR is given by:

$$CR = \frac{-t}{\ln\left(\frac{V}{V_0}\right)}$$

When you are dealing with exponential relationships, it is often useful to plot log–linear graphs, as shown in the example below.

Activity/Bq	Time/s
250	0
226	10
205	20
186	30
168	40
152	50
138	60
125	70

Example

Find the decay constant, and hence the half-life, of protactinium-234 from the data given. Assume that the data has already been corrected for background radiation.

The activity, A, at time t is given by $A = A_0 e^{-\lambda t}$.

Taking logarithms to the base e:

$\ln A = \ln(A_0 e^{-\lambda t})$

$\qquad = \ln A_0 + \ln e^{-\lambda t}$

$\qquad = \ln A_0 - \lambda t$

So plotting $\ln A$ on the y-axis against time t on the x-axis gives a straight line with a gradient of $-\lambda$.

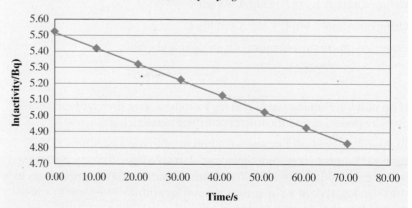

In(activity/Bq) against time/s

The gradient, $-\lambda$, is $-9.9 \times 10^{-3} \text{ s}^{-1}$.

$$\text{Half-life} = \frac{\ln 2}{\lambda}$$

$$= \frac{0.693}{9.9 \times 10^{-3} \text{ s}^{-1}}$$

$$= 70 \text{ s}$$

Data and formulae

FUNDAMENTAL CONSTANTS AND VALUES

Quantity	Symbol	Value	Units
speed of light in vacuo	c	3.00×10^8	m s^{-1}
permeability of free space	μ_0	$4\pi \times 10^{-7}$	H m^{-1}
permittivity of free space	ε_0	8.85×10^{-12}	F m^{-1}
charge of electron	e	-1.60×10^{-19}	C
the Planck constant	h	6.63×10^{-34}	J s
gravitational constant	G	6.67×10^{-11}	$\text{N m}^2 \text{ kg}^{-2}$
the Avogadro constant	N_A	6.02×10^{23}	mol^{-1}
molar gas constant	R	8.31	$\text{J K}^{-1} \text{mol}^{-1}$
the Boltzmann constant	k	1.38×10^{-23}	J K^{-1}
the Stefan constant	σ	5.67×10^{-8}	$\text{W m}^{-2} \text{K}^{-4}$
the Wien constant	α	2.90×10^{-3}	m K
electron rest mass (equivalent to $5.5 \times 10^{-4}\,\text{u}$)	m_e	9.11×10^{-31}	kg
electron charge–mass ratio	e/m_e	1.76×10^{11}	C kg^{-1}
proton rest mass (equivalent to $1.00728\,\text{u}$)	m_p	1.67×10^{-27}	kg
proton charge–mass ratio	e/m_p	9.58×10^7	C kg^{-1}
neutron rest mass (equivalent to $1.00867\,\text{u}$)	m_n	1.67×10^{-27}	kg
gravitational field strength	g	9.81	N kg^{-1}
acceleration due to gravity	g	9.81	m s^{-2}
atomic mass unit (1 u is equivalent to 931.3 MeV)	u	1.661×10^{-27}	kg

ASTRONOMICAL DATA

Body	Mass/kg	Mean radius/m
Sun	1.99×10^{30}	6.96×10^8
Earth	5.98×10^{24}	6.37×10^6

GEOMETRICAL EQUATIONS

arc length $= r\theta$

circumference of circle $= 2\pi r$

area of circle $= \pi r^2$

area of cylinder $= 2\pi rh$

volume of cylinder $= \pi r^2 h$

area of sphere $= 4\pi r^2$

volume of sphere $= \dfrac{4}{3}\pi r^3$

MOMENTUM

force
$$F = \frac{\Delta(mv)}{\Delta t}$$

impulse
$$F \Delta t = \Delta(mv)$$

CIRCULAR MOTION

angular velocity
$$\omega = \frac{v}{r}$$
$$\omega = 2\pi f$$

centripetal acceleration
$$a = \frac{v^2}{r} = \omega^2 r$$

centripetal force
$$F = \frac{mv^2}{r} = m\omega^2 r$$

OSCILLATIONS

acceleration
$$a = -(2\pi f)^2 x$$

displacement
$$x = A\cos(2\pi ft)$$

speed
$$v = \pm 2\pi f\sqrt{A^2 - x^2}$$

maximum speed
$$v_{\text{max}} = 2\pi fA$$

maximum acceleration
$$a_{\text{max}} = (2\pi f)^2 A$$

for a mass-spring system
$$T = 2\pi\sqrt{\frac{m}{k}}$$

for a simple pendulum
$$T = 2\pi\sqrt{\frac{l}{g}}$$

GRAVITATIONAL FIELDS

force between two masses
$$F = -\frac{G m_1 m_2}{r^2}$$

gravitational field strength
$$g = \frac{F}{m}$$

magnitude of gravitational field strength in radial field
$$g = \frac{GM}{r^2}$$

gravitational potential
$$\Delta W = m \Delta V$$
$$V = -\frac{GM}{r}$$
$$g = -\frac{\Delta V}{\Delta r}$$

ELECTRIC FIELDS AND CAPACITORS

force between two point charges
$$F = \frac{1}{4\pi\varepsilon_0}\frac{Q_1 Q_2}{r^2}$$

force on a charge
$$F = EQ$$

field strength for a uniform field
$$E = \frac{V}{d}$$

field strength for a radial field
$$E = \frac{Q}{4\pi\varepsilon_0 r^2}$$

electric potential
$$\Delta W = Q \Delta V$$
$$V = \frac{1}{4\pi\varepsilon_0}\frac{Q}{r}$$

capacitance
$$C = \frac{Q}{V}$$

decay of charge
$$Q = Q_0 e^{-t/RC}$$

time constant
$$RC$$

capacitor energy stored
$$E = \tfrac{1}{2}QV = \tfrac{1}{2}CV^2 = \tfrac{1}{2}\frac{Q^2}{C}$$

MAGNETIC FIELDS

force on a current
$$F = BIl$$

force on a moving charge
$$F = BQv$$

magnetic flux
$$\Phi = BA$$

magnetic flux linkage
$$\Phi = BAN$$

induced emf
$$\varepsilon = N\frac{\Delta\Phi}{\Delta t}$$

emf induced in a rotating coil
$$N\Phi = BAN\cos\theta$$
$$\varepsilon = BAN\omega\sin\omega t$$

transformer equations
$$\frac{N_s}{N_p} = \frac{V_s}{V_p}$$

efficiency
$$= \frac{I_s V_s}{I_p V_p}$$

RADIOACTIVITY AND NUCLEAR PHYSICS

the inverse square law for γ radiation
$$I = \frac{k}{x^2}$$

radioactive decay
$$\frac{\Delta N}{\Delta t} = -\lambda N$$
$$N = N_0 e^{-\lambda t}$$

activity
$$A = \lambda N$$

half life
$$T_{1/2} = \frac{\ln 2}{\lambda}$$

nuclear radius
$$R = r_0 A^{1/3}$$

energy–mass equation
$$E = mc^2$$

GASES AND THERMAL PHYSICS

gas law
$$pV = nRT$$
$$pV = NkT$$

kinetic theory model
$$pV = \tfrac{1}{3}Nm\left(c_{\text{rms}}\right)^2$$

kinetic energy of gas molecule
$$\tfrac{1}{2}m\left(c_{\text{rms}}\right)^2 = \tfrac{3}{2}kT = \frac{3RT}{2N_A}$$

energy to change temperature
$$Q = mc\,\Delta T$$

energy to change state
$$Q = ml$$

ASTROPHYSICS

1 astrophysics unit = 1.50×10^{11} m

1 light year = 9.45×10^{15} m

1 parsec = 206265 AU = 3.08×10^{16} m = 3.261 yr

Hubble constant $H = 65$ km s^{-1} Mpc^{-1}

lens equation $\qquad \dfrac{1}{f} = \dfrac{1}{u} + \dfrac{1}{v}$

$M = \dfrac{\text{angle subtended by image at eye}}{\text{angle subtended by object at unaided eye}}$

in normal adjustment $\qquad M = \dfrac{f_0}{f_e}$

resolving power $\qquad \theta \approx \dfrac{\lambda}{D}$

magnitude equation $\qquad m - M = 5 \log \dfrac{d}{10}$

Wien's law $\qquad \lambda_{\max} T = 0.0029$ m K

Hubble law $\qquad v = H d$

Stefan's law $\qquad P = \sigma A T^4$

Doppler shift for $v \ll c$ $\qquad z = \dfrac{\Delta f}{f} = -\dfrac{\Delta \lambda}{\lambda} = \dfrac{v}{c}$

Schwarzschild radius $\qquad R_s = \dfrac{2GM}{c^2}$

MEDICAL PHYSICS

lens equations $\qquad P = \dfrac{1}{f}$

$$m = \dfrac{v}{u}$$

$$\dfrac{1}{f} = \dfrac{1}{u} + \dfrac{1}{v}$$

intensity level $\qquad \text{intensity level} = 10 \log \dfrac{I}{I_0}$

absorption $\qquad I = I_0 e^{-\mu x}$

$$\mu_m = \dfrac{\mu}{\rho}$$

APPLIED PHYSICS

moment of inertia $\qquad I = \sum mr^2$

angular kinetic energy $\qquad E_k = \tfrac{1}{2} I \omega^2$

equations of angular motion $\quad \omega_2 = \omega_1 + \alpha t$

$$\omega_1^2 = \omega_1^2 + 2\alpha \theta$$

$$\theta = \omega_1 t + \tfrac{1}{2} \alpha t^2$$

torque $\qquad T = I \alpha$

angular momentum $\qquad \text{angular momentum} = I\omega$

work done $\qquad W = T\theta$

power $\qquad P = T\omega$

thermodynamics $\qquad Q = \Delta U + W$

$$W = p\Delta V$$

adiabatic change $\qquad pV^\gamma = \text{constant}$

isothermal change $\qquad pV = \text{constant}$

heat engines

efficiency $= \dfrac{W}{Q_{in}} = \dfrac{Q_{in} - Q_{out}}{Q_{in}}$

maximum efficiency $= \dfrac{T_H - T_C}{T_H}$

work done per cycle = area of loop

input power = calorific value × fuel flow rate

indicated power = (area of p-V loop) × (no. of cycles per second) × number of cylinders

output of brake power $P = T\omega$

friction power = indicated power – brake power

heat pumps and refrigerators

refrigerator: $COP_{\text{ref}} = \dfrac{Q_{\text{out}}}{W} = \dfrac{Q_{\text{out}}}{Q_{\text{in}} - Q_{\text{out}}}$

heat pump: $COP_{\text{hp}} = \dfrac{Q_{\text{in}}}{W} = \dfrac{Q_{\text{in}}}{Q_{\text{in}} - Q_{\text{out}}}$

TURNING POINTS IN PHYSICS

electrons in fields $\qquad F = \dfrac{eV_p}{d}$

$$F = Bev$$

$$r = \dfrac{mv}{Be}$$

$$\tfrac{1}{2} mv^2 = eV$$

$$\dfrac{QV}{d} = mg$$

$$F = 6\pi \eta r v$$

wave particle duality $\qquad c = \dfrac{1}{\sqrt{\mu_0 \varepsilon_0}}$

$$\lambda = \dfrac{h}{p} = \dfrac{h}{\sqrt{2meV}}$$

special relativity $\qquad E = mc^2 = \dfrac{m_0 c^2}{\left(1 - \dfrac{v^2}{c^2}\right)^{\frac{1}{2}}}$

$$l = l_0 \left(1 - \dfrac{v^2}{c^2}\right)^{\frac{1}{2}} \qquad t = t_0 \left(1 - \dfrac{v^2}{c^2}\right)^{-\frac{1}{2}}$$

Practice exam-style questions

1 Cassini-Huygens is an unmanned spacecraft which is studying Saturn and its moons. It was launched from Earth in 1997 and reached the Saturn moon, Titan, in 2005.

The spacecraft is powered by a radioactive thermal generator which uses heat generated by radioactive decay to generate electricity. The radioisotope plutonium-238 is used.

The table shows the properties of three isotopes of plutonium.

	Pu-238	Pu-239	Pu-241
Half-life (in years)	87.74	24 110	14.4
Specific activity (GBq/gram)	640	0.063	104
Principal decay mode	alpha	alpha	beta
Decay energy (MeV)	5.593	5.244	0.021
Radiological hazards	alpha, weak gamma	alpha, weak gamma	beta, weak gamma

(a) What is meant by the **activity** of a radioactive source?

_____

1 mark

(b) Show that a power of 573 W is generated by the alpha emissions of 1 kg of plutonium-238.
    (Use data from the table above.)

_____

_____

_____

3 marks

(c) The electricity generation process is 6.7% efficent. What mass of plutonium-238 is needed to generate 600 W of electrical power?

_____

_____

_____

3 marks

(d) The Cassini mission was planned to last for 11 years. If the spacecraft needed 600 W at the end of the mission, what mass of plutonium-238 was needed at the beginning of the mission?

_____

_____

_____

3 marks

(e) Explain why plutonium-238 was preferred to the other isotopes shown in the table above.

_____

_____

_____

2 marks

(f) There were a number of protests against the Cassini mission, principally due to the use of radioactive power supplies. Explain why the protestors may have been concerned.

_____

_____

_____

2 marks

(g) Suggest why a radioactive power source was preferred to using solar panels to power the spacecraft.

_____

_____

_____

2 marks

Total Marks: 16

2 During 1 hour of exercise in a gym, a person may lose 500 g of water through sweating. Sweating is a vital part of the body's mechanism for maintain a steady temperature. It is important to replace the lost water by drinking water.

The latent heat of vaporisation of sweat is $2.4 \times 10^6 \, \text{J kg}^{-1}$.

(a) Explain what is meant by the **latent heat of vaporisation** of a liquid.

_____

_____

2 marks

(b) Explain how sweating helps to keep a person cool.

_____

_____

_____

2 marks

(c) Calculate the average rate of heat loss through sweating for an hour's exercise in the gym.

_____

_____

_____

3 marks

Total Marks: 7

3 In October 2009, a large helium balloon belonging to amateur scientist, Richard Heene, floated off from his garden in Colorado, USA. At the same time, Heene's 6-year-old son went missing. Panic ensued until the boy was found safe and sound a few hours later.

(a) The volume of the balloon was $28.3 \, \text{m}^3$. If the temperature was $20 \, °\text{C}$ and atmospheric pressure was $1.01 \times 10^5 \, \text{Pa}$, calculate the mass of helium in the balloon. (The molar mass of helium is $4.0 \times 10^{-3} \, \text{kg}$.)

_____

_____

_____

3 marks

(b) Suppose the balloon rose to a height of 5 km, where the air pressure was 0.5×10^5 Pa and the temperature was $-40\,°C$. Calculate the new volume of the balloon. State any assumptions that you have made.

_____

_____

_____

3 marks

(c) Any helium balloon has a maximum height that it can reach. Suggest why there is a maximum height.

_____

_____

1 mark

Total Marks: 7

4 **(a)** Explain what is meant by the **mass difference** and the **binding energy** of a nucleus.

(i) Mass difference _____

_____

(ii) Binding energy _____

_____

2 marks

(b) The graph below shows how the binding energy per nucleon varies with nucleon number.

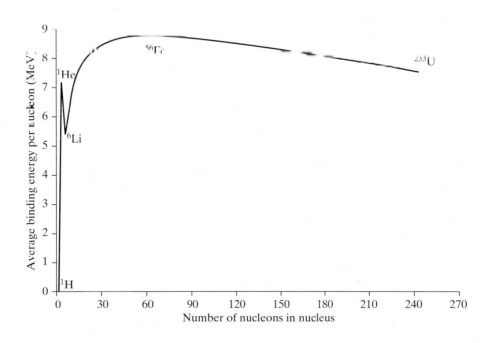

(i) The mass of helium-4 is 4.00151 u. Show that the binding energy per nucleon for ^4_2He plotted on the graph, 7.1 MeV, is correct. (You will need to use other values from the data sheet.)

_____

_____

_____

_____

3 marks

(ii) Use the graph to explain why energy is released by the nuclear fission of a uranium-235 nucleus.

_____

_____

_____

2 marks

(c) In a nuclear reactor, uranium-235 absorbs a neutron to become uranium-236, which decays by fission:

$$^{235}_{92}\text{U} + ^1_0\text{n} \rightarrow ^{236}_{92}\text{U} \rightarrow ^{94}_{40}\text{Zr} + ^{140}_{56}\text{Ba} + 2^1_0\text{n} \ (+ \ \text{beta particles})$$

Calculate the energy released by this fission reaction. Give your answer in joules.

Mass of U-235 = 235.044 u

Mass of Zr-94 = 93.906 u

Mass of Ba-56 = 139.91 u

(See the data sheet for other values needed.)

_____

_____

_____

_____

3 marks

Total Marks: 10

Answers, explanations, hints and tips

Question	Answer		Marks
1 (a)	Activity is the number of emissions per second.	(1)	1
1 (b)	Activity of 1 kg is 640 GBq $\times$ 1000 = 640 $\times$ 10^{12} Bq.	(1)	
	Each alpha particle is emitted with energy		
	5.593 MeV = 5.593 $\times$ 10^6 $\times$ 1.6 $\times$ 10^{-19} J		
	$\qquad$ = 8.95 $\times$ 10^{-13} J	(1)	
	So power (= energy/sec) = 8.95 $\times$ 10^{-13} J $\times$ 640 $\times$ 10^{12} s^{-1}		
	$\qquad$ = 573 W.	(1)	3
1 (c)	1 kg generates 573 W, but electricity generation only 6.7%		
	efficient so electrical power = 0.067 $\times$ 573 = 38.4 W.	(1)	
	So 1 W of electrical power needs 1000/38.4 = 26 g plutonium.	(1)	
	And 600 W needs 15.6 kg.	(1)	3
1 (d)	Since $N = N_0\,e^{-\lambda t}$ and mass $m \propto N$,		
	$m = m_0\,e^{-\lambda t}$	(1)	
	$\lambda = \ln 2/T_{1/2} = 0.693/87.74 = 7.9 \times 10^{-3}$ years^{-1}		
	$\lambda t = 7.9 \times 10^{-3}$ years^{-1} $\times$ 11 years = 0.087	(1)	
	$m_0 = m\,e^{\lambda t} = 15.6\,e^{0.087} = 17$ kg	(1)	3
1 (e)	Longer half-life needed for long mission into space.		
	Alpha emitter so easier to shield.		
	Alpha emitter so all energy transferred close to power source.		
	High power density/lots of energy.		
	(Any 2 points)	(2)	2
1 (f)	Danger of radioactive/plutonium contamination if rocket		
	blew up/crashed.	(1)	
	Radioactive materials can cause increased cancer risk/		
	plutonium is toxic.	(1)	2
1 (g)	Sun's energy is too dilute/inverse-square law for Sun's radiation		
	means that solar radiation not intense enough at the distance		
	of Saturn	(1)	
	Solar panels would have to be very large/too heavy for lift-off.	(1)	2
			Total 16

Question	Answer		Marks
2 (a)	The energy required to vaporise 1 kg of liquid	(1)	
	with no increase in temperature	(1)	2
2 (b)	Water in sweat needs energy to evaporate/latent heat		
	of vaporisation.	(1)	
	This energy comes from the body's heat, causing it to cool down.	(1)	2
2 (c)	Heat loss in 1 hour $Q = ml$	(1)	
	$= 0.5 \times 2.4 \times 10^6 = 1.2 \times 10^6$ J	(1)	
	Average rate of heat loss $= 1.2 \times 10^6$ J/3600 s $= 330$ W.	(1)	3
			Total 7
3 (a)	$pV = nRT$, so $n = PV/RT = (1.01 \times 10^5 \times 28.3)/(8.31 \times 293)$	(1)	
	$= 1170$ moles	(1)	
	Since 1 mole of helium has a mass of 4×10^{-3} kg, the balloon		
	contains $1174 \times 4 \times 10^{-3} = 4.7$ kg of helium.	(1)	3
3 (b)	pV/T is constant, so $V_2 = p_1V_1T_2/T_1p_2$	(1)	
	$= (1.01 \times 10^5 \times 28.3 \times 233)/(293 \times 0.5 \times 10^5) = 46\,\text{m}^3$	(1)	
	This assumes that the gas behaves as an ideal gas/		
	no helium escapes.	(1)	3
3 (c)	Pressure of atmosphere gets less with height, so helium		
	balloon keeps expanding until it bursts.	(1)	1
			Total 7
4 (a)	**(i)** Mass difference is the difference between the mass of a		
	nucleus and the total mass of its constituent nucleons		
	(protons and neutrons).	(1)	
	(ii) Binding energy is the energy required to split a nucleus		
	into its constituent nucleons, or the energy released when the		
	nucleus is formed from its constituents.	(1)	2
4 (b)	**(i)** Mass difference $= 2m_\text{p} + 2m_\text{n} - 4.00151$		
	$= (2 \times 1.00728) + (2 \times 1.00867) - 4.00151 = 0.030$ u	(1)	
	Binding energy $= 0.030 \times 931.3$ MeV $= 28.3$ MeV	(1)	
	Binding energy per nucleon $= 28.3/4 = 7.08$ MeV/nucleon	(1)	3
	(ii) U-235 has binding energy per nucleon of about		
	7.5 MeV/nucleon. Fission produces two smaller nuclei;		
	these have a larger binding energy per nucleon		
	(about 8.5 MeV/nucleon).	(1)	
	This means there is an overall mass loss in the fission, and so		
	energy is released (as kinetic energy of the fission products).	(1)	2
4 (c)	Original mass $= 235.044$ u $+ 1.009$ u $= 236.053$ u		
	Final mass $= 93.906$ u $+ 139.91$ u $+ (2 \times 1.009$ u$) = 235.836$ u	(1)	
	Mass difference $= 0.217$ u, giving an energy of		
	$0.217 \times 931.3 = 202$ MeV	(1)	
	$= 202 \times 10^6 \times 1.6 \times 10^{-19} = 3.23 \times 10^{11}$ J	910	3
			Total 10

Option Unit 5A Astrophysics

A.1.1 Lenses and optical telescopes

Lenses

Galileo was the first astronomer to use a telescope, some 400 years ago. This not only revolutionised our understanding of our Solar System but subsequently of the Universe as a whole. The telescope he used was based on lenses made of glass, which alter the direction of light rays.

There are two basic types of optical lens. A **concave lens**, also called a **diverging lens**, spreads an incident beam of light into a diverging emergent beam. A **convex lens**, also called a **converging lens**, can focus an incident beam. Lenses are used in optical instruments such as binoculars, slide projectors, cameras, spectacles, magnifying glasses and telescopes to produce an image.

For a single converging lens, the line that passes through the centre of the lens at right angles to it is called the **principal** or **optical axis**. Light rays from a distant object that are essentially parallel to the principal axis of the lens converge to a point called the **principal focus**, F (Fig 1). The distance between the principal focus and the centre of the lens is called the **focal length**, f. The shorter the focal length of a converging lens, the more strongly it converges light rays.

Fig 1
A ray diagram showing the action of a converging lens on a beam of light

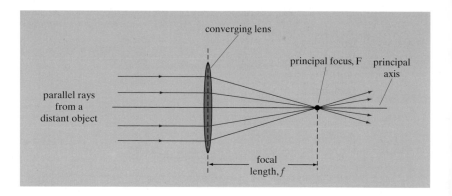

The construction of a **ray diagram** is the best method to gain a good visual understanding of the way an incident light beam behaves on passing through a lens system. The object distance (measured along the principal axis from the centre of the lens) is denoted by u and the corresponding image distance by v. A converging lens can produce both real and virtual images.

- When an object is further away from the lens than the focal length ($u > f$), a real image is formed, inverted, on the far side of the lens (Fig 2). A real image can be formed on a screen.

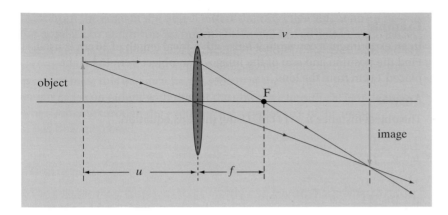

Fig 2
A ray diagram for a converging lens, showing how a real image is formed

Essential Notes

When drawing ray diagrams remember that:

- light rays that pass through the centre of the lens are undeviated

- light rays parallel to the principal axis converge to the focal point.

- When an object is closer to the lens than its focal length ($u < f$), a virtual image is formed on the same side of the lens as the object, not inverted (Fig 3). A virtual image cannot be captured on a screen – think of a magnifying glass.

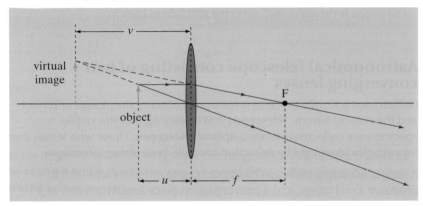

Fig 3
A ray diagram showing a converging lens producing a virtual image

Essential Notes

In reality light rays change direction at each surface of the lens but by convention ray diagrams show the rays changing direction just once.

For a single lens the relationship between the object distance, u, the image distance, v, and the focal length, f, is given by the lens equation:

$$\frac{1}{u} + \frac{1}{v} = \frac{1}{f}$$

The sign convention for this equation is important:

- f is positive for a converging lens and negative for a diverging lens

- u is positive for a real object

- v is positive for a real image but negative for a virtual image.

The **magnification**, M, of a lens is given by the ratio of the image distance to object distance:

$$M = \frac{v}{u}$$

The same sign convention must be used here, so:

- a negative magnification shows that, for a real object, the image is virtual

- a positive magnification shows that, for a real object, the image is real and hence inverted.

Examiners' Notes

You don't need to remember the lens equation; it will be provided. But you do need to remember the sign convention.

Fig 5
Chromatic aberration causes light of different wavelengths (colours) to focus at different positions along the optical axis

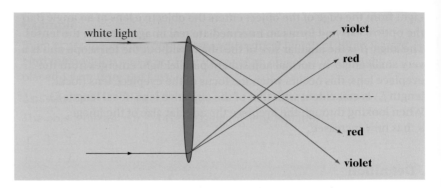

Fig 6
Spherical aberration causes rays to focus at different positions, causing image blurring

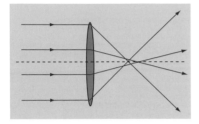

Examiners' Notes

In ray diagrams depicting aberrations it is important to show a minimum of four rays parallel to the principal axis.

Spherical aberration results in light rays in a parallel beam being focused at slightly different positions (Fig 6). This is because of the curvature of the lens (or mirror). Light rays near the edge of the lens are deviated more than those near the optical axis. The effect is most pronounced in lenses (or mirrors) of large diameter, resulting in a blurring of the image. The effect can be minimised by making both lens surfaces contribute equally to the ray deviations.

Reflecting telescopes

Reflecting telescopes use curved mirrors to focus the parallel light from a distant object. The ideal curved mirror is parabolic in shape as this focuses the parallel light rays from a distant object to a single focal point, eliminating spherical aberration. The mirror itself consists of a very thin coat of silver or aluminium atoms that have been deposited onto a backing material. The thickness of this coat is often less than 25 nm (2.5×10^{-10} m) to provide as smooth a surface as possible and so minimise possible distortions.

There are two principal ways that the image can be viewed. The **Newtonian telescope** uses a flat secondary mirror which reflects rays from the curved primary mirror out of the side of the telescope, where they are focused at an eyepiece or electronic camera (Fig 7).

Fig 7
The Newtonian arrangement for a reflecting telescope

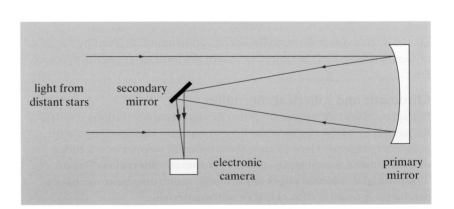

The **Cassegrain telescope** uses a convex secondary mirror that sends the rays down an opening in the primary mirror (usually in the centre) to a focus at an eyepiece or camera (Fig 8).

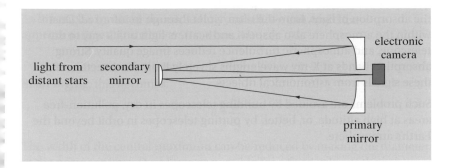

Fig 8
The Cassegrain arrangement for a reflecting telescope

This Cassegrain arrangement is used in most modern optical telescopes. The main advantage of this arrangement is that large, heavy and often complex equipment can be attached to the base of the telescope.

The diameter of the primary mirror determines the ability of the telescope to collect light. The light-gathering power is proportional to the square of the diameter of its primary mirror. Modern reflecting telescopes use mirrors up to 10 m in diameter.

Chromatic aberration does not affect reflecting telescopes. Table 1 lists the disadvantages of refracting telescopes and the relative advantages of reflecting telescopes.

Disadvantages of refracting telescopes	Advantages of reflecting telescopes
• Mounting of the lens and support can only be made using the edge of the lens	• Large single mirrors can be made which are light and easily supportable from behind
• Using glass of sufficient clarity and purity and free from defects to make large-diameter telescopes is extremely difficult	• Mirror surfaces can be made just a few nanometres thick, giving excellent image properties
• Large-diameter lenses are heavy and tend to distort under their own weight	• Mirrors use only the front surface for reflection so removing many of the problems associated with lenses
• Suffer from chromatic aberration and spherical aberration	
• Heavy and difficult to manoeuvre quickly	• No chromatic aberration, and no spherical aberration when using parabolic mirrors
• Difficult to mount heavy observing equipment and associated electronics	• Relatively light mirrors allow rapid response to astronomical events
• Large magnifications require large objective lenses and very long focal lengths	• Smaller segmented mirrors can be used to form a large composite objective mirror

Table 1
Comparing refracting and reflecting telescopes

Example

The diameter of the objective mirror of the Hale telescope is 5.1 m, and it is observing a star emitting light of wavelength 510 nm.

(a) What is the angular resolution of the telescope in radians?

(b) What is the smallest detail the telescope can detect on the surface of the Moon?

(Distance to Moon = 3.8×10^8 m; take the wavelength of light detected to be 510 nm.)

Answer

(a)

$$\theta = \frac{510 \times 10^{-9}}{5.1} = 1.0 \times 10^{-7} \text{ radians}$$

(b) The smallest feature on the Moon would be determined by the value x, given by

$$x = L\theta = 3.8 \times 10^8 \times 1.0 \times 10^{-7} \approx 40 \text{ m}$$

Charge coupled devices

Many telescopes have now had photographic film replaced with electronic detectors known as **charge coupled devices** or **CCDs**. These are basically a type of microchip that converts a light signal directly into a digital format. They are made from thin wafers of silicon divided into individual picture elements called **pixels**. A typical CCD may have many millions of pixels covering an area of only a few cm^2. Photons striking a pixel cause electrons to be liberated from the atoms on the silicon surface.

CCDs have important advantages over photographic film:

1. They are very efficient. The **quantum efficiency** of a CCD is the ratio of the number of photons detected to the number of photons incident. In modern devices this is >75% (compared to ~5% for photographic film). Objects some 10 to 20 times fainter can be imaged, or shorter exposure times used.

2. The image is in digital format. The liberated electrons are trapped by positive electrodes and an electric charge is built up in each pixel. The amount of charge is directly proportional to the number of incident photons, i.e. to the brightness of the incident light (a linear response). On completion of the exposure, the charge on each pixel is read out digitally for direct processing by computer and further image analysis.

3. Because of the way CCDs record the information, some pixels may contain only a few electrons while others have several thousands of electrons; modern devices can readily handle differences in signal intensity of 20 orders of magnitude.

4. CCDs can detect wavelengths beyond those of visible light, extending from low-energy X-rays to the infra-red, and this has opened up a vast new area for astronomers to study.

Essential Notes

The large working range of CCDs makes them ideal for imaging objects that vary in brightness by tens of thousands of times, a difference that often occurs in galaxies between the bright nucleus and the faint spiral arms.

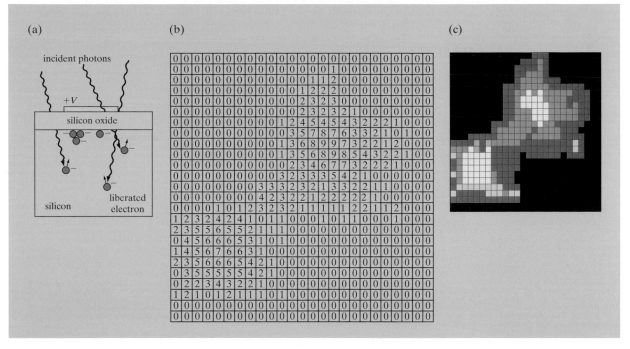

Fig 11
(a) Photons incident on the CCD create an electric charge which is collected to form a digital image; (b) a simplified example of data from a CCD array with numbers from 0 to 9 representing the intensity of radiation striking that particular pixel; (c) intensity levels on a computer screen provide an image

A.1.2 Non-optical telescopes

Single dish radio telescopes

The development of radio astronomy is recent compared with optical astronomy, but has undergone considerable advances since the early pioneering days in the 1930s. Radio astronomy uses a window between 30 MHz and 600 GHz, corresponding to wavelengths of 10 m and 0.5 mm, respectively. Radio waves are emitted by the Sun, interstellar gas clouds, nebulae and supernova remnants, as well as by new classes of objects that would otherwise be completely unknown. Perhaps the major advantage over optical astronomy is the ability to record data continuously 24 hours a day.

The operation of a **radio telescope** is basically the same as that of an optical reflector (Fig 8); the radio waves are collected by a large parabolic dish and brought to a focus. The steerable dish scans back and forth across the source of radio waves, and the signals are amplified and computer-analysed to produce a visual image. The intensity of the focused signal is proportional to the collecting power of the telescope; the larger the dish, the larger the collecting power: collecting power $\propto$ diameter^2. The large diameter also provides the greatest resolving power (refer to the Rayleigh criterion, page 71) but, because of the long wavelengths involved, radio telescopes have poor resolving capabilities compared with their optical counterparts.

Essential Notes

$1 \text{ minute of arc} (1') = \left(\frac{1}{60}\right)^\circ$

$1 \text{ second of arc} (1'') = \left(\frac{1}{3600}\right)^\circ$

The world's largest single dish steerable radio telescope, the National Radio Astronomy Observatory in West Virginia, is 105 m in diameter and can achieve a resolution of about 1 minute of arc at a wavelength of 3 cm.

> ### Example
>
> Compare the resolving powers of a radio telescope with a dish size of 5.1 m detecting radio signals with a wavelength of 2 m, and the Hale optical telescope with a mirror of 5.1 m collecting light at a wavelength of 510 nm.
>
> ### Answer
>
> Using the expression for image resolution,
>
> $$\theta \approx \frac{\lambda}{D}$$
>
> Radio telescope gives
>
> $$\theta \approx \frac{2}{5.1} = 0.39 \text{ radians}$$
>
> Optical telescope gives
>
> $$\theta \approx \frac{510 \times 10^{-9}}{5.1} = 1.0 \times 10^{-7} \text{ radians}$$
>
> i.e. a factor of almost 4×10^6 in favour of the optical telescope.

Poor angular resolution of radio telescopes can be overcome by connecting two or more together to form a system that is better than the best optical telescopes, a technique known as interferometry. Resolutions in the milliarcsecond (0.001″) range can be achieved with this system.

Radio telescopes are not immune from interference. Below the lower band limit of 30 MHz the ionosphere itself strongly absorbs the signal, while above 60 GHz absorption by water vapour in the atmosphere is a significant problem. Between these frequencies, man-made interferences, such as those produced from mobile phones, radio telephones and radar scanners, can pose serious problems with sensitive instrumentation and so radio telescopes tend to be located in isolated areas.

I-R, U-V and X-ray telescopes

Infra-red (I-R) astronomy

Infra-red astronomy is used to make observations of cool regions (temperatures between a few tens to a hundred kelvin) such as interstellar gas, cooler stars, star formation regions and active galaxies, and of the large-scale structure of the Universe. Most infra-red radiation is absorbed in the Earth's atmosphere, principally by carbon dioxide and water vapour, but a few windows are available that allow ground-based I-R telescopes to make useful observations. The largest I-R telescopes are the 3.8 m UK Infra-Red Telescope (UKIRT) and the 8.2 m Subaru Telescope, both based at the Mauna Kea Observatory in Hawaii. In these telescopes the infra-red radiation is detected using a bolometer (a device that measures subtle changes in temperature), cooled to liquid nitrogen temperatures and shielded from other sources of infra-red radiation including its own emissions.

Space telescopes offer significant improvements. The IRAS (Infra-Red Astronomy Satellite), ISO (Infra-red Space Observatory) and the Spitzer Space Telescope (SST) have allowed much more detailed infra-red observations, particularly in active galaxies and quasars. Infra-red radiation is not scattered by particles in the dusty regions as much as shorter-wavelength visible light, but passes through, revealing distant stars and objects that are hidden at visible wavelengths.

Ultra-violet (U-V) telescopes

The Earth's atmosphere is totally opaque to ultra-violet radiation below 300 nm, hence most ultra-violet astronomy using ground-based observatories has been undertaken between 310 and 400 nm. The more important region between 10 and 310 nm can only be observed from above the atmosphere by space telescopes. The International Ultra-violet Explorer (IUE) operated for 18 years collecting data on the composition of cometary tails and on the energy profiles of exploding stars. The extreme ultra-violet region (XUV) has now been studied with the ROSAT and EUVE telescopes, which have also conducted whole-sky surveys. Major sources of ultra-violet radiation include hot, massive stars, newly formed white dwarfs and the core regions of active galaxies.

X-ray telescopes

X-ray astronomy involves radiation imaging at X-ray and gamma-ray wavelengths. The first X-ray source, Scorpius X-1, was discovered in 1962. Since then dedicated satellites, such as the Einstein Observatory and the XMM-Newton Observatory, have been used to undertake whole-sky surveys, revealing hundreds of thousands of cosmic X-ray sources. X-rays come from extremely hot gas in the range 10^6–10^8 K normally associated with highly energetic processes, and as such provide a rich array of objects to study, including interacting binary stars, active galaxies, galaxy clusters and supernova remnants. Interest has also focused on pulsars, neutron stars and black holes. The telescope INTEGRAL is currently providing data on hundreds of sources both in our own galaxy and in the distant Universe. Future space-borne telescopes are likely to operate in the hard (short-wavelength) X-ray region.

Gamma-ray telescopes extend observations to wavelengths shorter than 0.01 nm. The space-borne Compton Gamma Ray Observatory (CGRO) operated between 1991 and 2000 and observed over 300 gamma-ray sources each year. The major sources of such high-energy radiation include solar flares, pulsars, quasars, active galaxies and supernova remnants. Sudden bursts of gamma radiation that last from 0.01 s to 1000 s have also been detected in all parts of the sky. The origin of these gamma-ray bursts (GRBs) is unknown and there is a very active research programme to study these.

Classification of stars

Classification by luminosity

Essential Notes

Newton's Law of Gravitation and Coulomb's Law are just two examples of the inverse square law. See Unit 4.

The observed brightness of a star depends on how far away it is; the light received obeys the inverse square law.

> **Definition**
>
> *The inverse square law states that the magnitude of a physical property is proportional to the reciprocal of the square of the distance from a source.*

The **luminosity**, L, of a star is the total amount of energy it radiates per second, i.e. its power (measured in joules per second or watts). The light received from the star obeys the inverse square law; if we assume that the star is a point source of radiation a distance R away then the observed **brightness**, b (sometimes called intensity, or radiation flux), is the amount of energy radiated per second per square metre over the surface of a sphere of radius R:

$$b = \frac{L}{4\pi R^2}$$

Brightness has the unit W m^{-2}.

Apparent magnitude

Essential Notes

The magnitude scale is an example of a logarithmic scale.

When astrophysicists discuss a star's brightness it is usually expressed on a scale that has developed historically and is called the magnitude scale. The ancient Greek astronomers Hipparchus and Ptolemy divided visible stars into six magnitude classes, with magnitude $+1.0$ representing the brightest stars and magnitude $+6.0$ representing those just visible to the naked eye, i.e. the faintest.

Because this classification of magnitude took no account of distance, we now refer to this as a star's **apparent magnitude**, m. Using this scale, stars of apparent magnitude $+1.0$ are about 100 times as bright as stars of apparent magnitude $+6.0$. This means that the difference of 5 magnitudes corresponds to a brightness ratio of 100, and therefore a change of 1 magnitude corresponds to a brightness ratio of $100^{1/5} = 2.5$.

Essential Notes

Here 'log' means log to the base 10, i.e. $\log_{10}$.

The subjective scale for brightness can be quantified as:

$$m = -2.5 \log b$$

Examiners' Notes

Note the lower case m for apparent magnitude.

where m is the apparent magnitude and b is the star's corresponding brightness.

The scaling factor of 2.5 ensures that a change of 5 magnitudes corresponds to a brightness ratio of 100 as discussed above.

The introduction of the telescope has led to many stars being assigned apparent magnitudes which are outside the range of magnitude $+1.0$ to magnitude +6.0. Barnard's star is +9.5 and large telescopes can detect stars as faint as +21.0. Negative values are needed to denote very bright objects. The full Moon has an apparent magnitude of –12.7 and the Sun is −26.74.

Definition

When comparing two stars, the difference in their apparent magnitudes can be given by the expression:

$$m_2 - m_1 = -2.5 \log\left(\frac{b_2}{b_1}\right)$$

where m_1, m_2 are the apparent magnitudes of stars 1 and 2, with b_1, b_2 their respective brightnesses.

Essential Notes

- The more negative the value of the apparent magnitude, the brighter the star appears.

- The larger the magnitude, the fainter the star appears.

Examiners' Notes

You need to remember this relationship.

Absolute magnitude

For direct comparisons between stars it is more meaningful to eliminate varying distances and base the comparison on the brightness at a fixed distance from the Earth. This standard distance is taken as 10 **parsec** or 32.6 light years. The magnitude at this distance is called the **absolute magnitude**, *M*. Any difference in absolute magnitude is a result of difference in luminosity. As with apparent magnitude, negative values indicate very bright stars.

The distances to nearby stars can be found by trigonometric **parallax**. This technique relies on the fact that as the Earth orbits the Sun, the apparent position of a nearby star changes relative to the more distant background stars and galaxies. The angle of parallax is the angle subtended at the star by the Earth–Sun distance (Fig 12), and although this angle is very small it can be measured by the displacement of the star over six months.

Examiners' Notes

Note the upper case M for absolute magnitude.

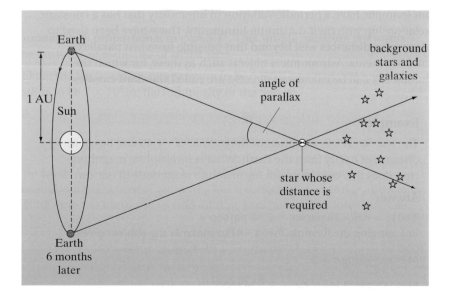

Fig 12
Measurement of the angle of parallax allows the distance to the star to be calculated

Example

The Sun has a surface temperature of 5800 K and a radius of 7×10^8 m. What would be the radius of a star that would produce the same output power of the Sun if it surface temperature was only half that of the Sun?

Answer

Using $P = \sigma A T^4$ for both the star and the Sun allows us to write

$$\sigma A_{Sun} T_{Sun}{}^4 = \sigma A_{star} T_{star}{}^4$$

which leads to

$$r_{star}^2 = r_{Sun}^2 \frac{T_{Sun}^4}{T_{star}^4}$$

$$= (7 \times 10^8)^2 \times 2^4 = 7.8 \times 10^{18}$$

and so

$$r_{star} = 2.8 \times 10^9 \text{ m}$$

i.e. some 4 times greater in radius.

Black-body radiation

An object that absorbs all the radiation that falls on it and reflects none is called a **black body**. If such a body is in thermal equilibrium with its surroundings then it must emit radiation at the same rate as it absorbs it. A perfect black body is therefore an ideal radiator.

Hot objects such as stars are efficient emitters of electromagnetic radiation; they effectively emit **black-body radiation**, the intensity of which at a particular wavelength depends only on the surface temperature of the object. Think of a metal being heated; initially it glows dull red, and as the temperature increases the colour changes from bright red to orange, then to yellow and eventually to a bluish white. These are similar to observed star colours. The wavelength of light that is predominantly emitted becomes shorter as the temperature is increased. The pattern of power radiated against wavelength for a particular temperature is called a **black-body curve** (Fig 14).

A key feature of the black-body curves is the shift of the peak wavelength, λ_{max} (the wavelength at which the radiation shows maximum intensity) as the temperature falls. The relationship between λ_{max} (in metres) and T (in kelvin) is known as **Wien's Displacement Law** (or just Wien's Law) and is given by:

$$\lambda_{max} T = \text{constant} = 2.9 \times 10^{-3} \text{ m K}$$

The peak wavelength is inversely proportional to the temperature, so the dominant wavelength of a black-body radiator decreases as it gets hotter.

A star with a temperature of several million degrees will emit radiation in the X-ray region of the spectrum, whereas an object at room temperature,

Essential Notes

Black-body radiation is covered in Collins Student Support Materials, Unit 1, section 3.1.1

Examiners' Notes

The units for the constant here are metre kelvin (m K), **not** to be confused with millikelvin (mK).

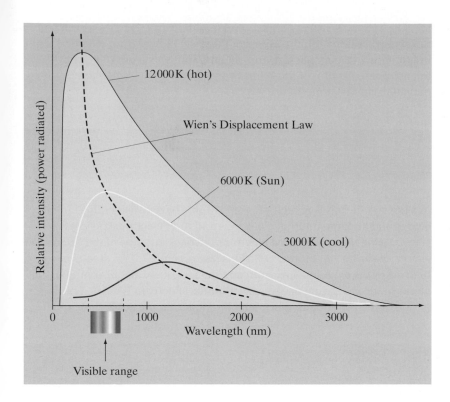

Fig 14
Black-body radiation curves (also called Planck curves) at selected temperatures. The area under each of the curves is equivalent to the total power as given by Stefan's Law. The wavelength at the peak intensity is governed by Wien's Displacement Law

say 300 K, emits radiation mainly in the infra-red. An object near absolute zero on the other hand would emit radiation primarily as microwaves (see Cosmic microwave background, page 95).

The direct connection between the intensity distribution of radiation and the temperature provides a method for determining the surface temperatures of stars. Measurements of the intensity distribution of the Sun over a broad range of wavelengths clearly shows the Sun to be a good approximation to a black body with $\lambda_{max} = 500\,\text{nm}$ or $500 \times 10^{-9}\,\text{m}$. Using Wien's Displacement Law this gives a surface temperature of 5800 K.

Principles of the use of stellar spectral classes

As we saw above, the colour of a star is a good guide to its approximate surface temperature, because its black-body radiation spectrum depends only on its surface temperature; dull red stars are cool and bluish-white stars are very hot. The radiation detected on Earth from a star can tell us much more than this. **Stellar spectroscopy** has become a powerful tool in determining not only the precise surface temperature but the composition of and physical conditions within stars. Spectroscopy gives rise to three types of spectra: an emission or bright line spectrum, a continuous spectrum, and an absorption spectrum. Each of these gives different information about its source.

In a gas at low temperature and pressure, almost all the atomic electrons are in the lowest energy level (ground state). As the temperature increases, more atomic collisions take place and electrons are raised to excited states. These electrons eventually return to lower energy levels, emitting photons

Essential Notes

Stellar spectra can be observed when diffraction gratings are used together with optical telescopes. Diffraction gratings are dealt with in Unit 2.

at precise characteristic energies corresponding exactly to the spacing of the energy levels within the atoms of the gas. The spectrum recorded is of bright lines on a dark background, an **emission spectrum** (Fig 15), with the intensity and position of these lines corresponding to particular electronic transitions in the atoms of the gas.

Fig 15
An emission spectrum (for hydrogen) showing bright lines on a dark continuous background

In a hot star the gas is at high pressure. Atoms have considerable kinetic energy and undergo multiple collisions and their electrons are in excited states. By the time the excited electrons fall back into one of the discrete energy levels, further atomic collisions have occurred. This results in a blurring of the emission spectrum and the loss of any detail about the atoms in the gas, giving rise to a **continuous spectrum** (Fig 16). This is typical of the continuous spectrum obtained from the Sun's photosphere.

Fig 16
A continuous spectrum showing no line structure

Essential Notes

The dark lines in an absorption spectrum match exactly the bright lines in an emission spectrum produced by the same gas.

The photosphere acts as a source of white light. This light then passes through the outer regions of the Sun, which are a lot cooler and composed of mainly hydrogen gas. Photons of the characteristic energies of the transitions in the gas will be absorbed and electrons raised to an excited state (perhaps to the second level or shell, $n = 2$, or even higher shells, $n = 3, 4, 5, 6$ and so on). As electrons falls back to the first level, $n = 1$ (the ground state), or intermediate levels, photons are emitted but in random directions. The resulting spectrum comprises dark lines (undetected photons) characteristic of an **absorption spectrum**.

Fig 17
An absorption spectrum (for hydrogen) showing dark lines on a continuous background

Essential Notes

Energy levels and line spectra are covered in Unit 1.

The absorption lines for hydrogen in the visible part of the spectrum result from electrons moving from the first excitation level ($n = 2$) to higher energy levels. This leads to a series of lines shown in Fig 17, called the **Balmer series**. The intensity of the Balmer lines depends on the particular temperature of the star. Other dark lines in an absorption spectrum, characteristic of other particular elements within the gas, help to determine the temperature of the star accurately (Fig 18).

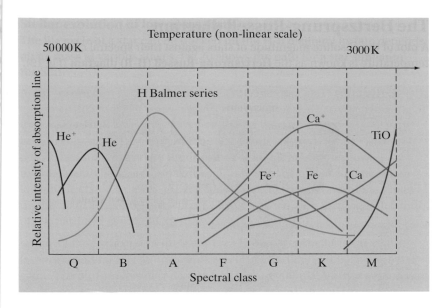

Fig 18
The intensity of particular absorption lines as a function of the temperature, and hence stellar spectral class. The Balmer series is a particularly useful indicator covering all spectral classes

A full analysis of a star's absorption spectrum reveals not only what particular materials are contained within the star, but can also indicate the state of the atoms, i.e. neutral or ionised. The relative strength of particular absorption lines gives the **spectral class** of a star. These are designated by one of seven letters:

O, B, A, F, G, K, M

O is the hottest type and M the coolest type, but the relationship between temperature and spectral class is neither linear nor logarithmic.

Table 2 shows the properties of stars in these seven spectral classes.

Spectral class	Colour	Temperature range (K)	Prominent absorption lines	Example star
O	blue, U-V	25 000 – 50 000	H, He, He$^+$	10 Lacertae
B	blue, U-V	11 000 – 25 000	H (Balmer lines), He	Rigel, Spica
A	bluish white	7500 – 11 000	H, ionised metals	Sirius, Vega
F	white	6000 – 7500	ionised metals: Ca$^+$, Fe$^+$	Procyon
G	yellow-white	5000 – 6000	ionised Ca$^+$, neutral metals	Sun, Capella
K	orange	3500 – 5000	neutral metals: Fe	Aldebaran
M	red	<3500	neutral compounds, TiO	Betelegeuse, Antares

Table 2
The classification of stars by spectral class, indicating the temperature range and prominent absorption lines

In practice the spectral classes are more complex and further subdivided into ten divisions from 0 to 9. By convention, the lower the number the hotter the star. A star like our Sun is designated a G2 star; this is hotter than a G3 star but cooler than a G1 star.

Essential Notes

The mnemonic

Oh **B**e **A** **F**ine **G**irl, **K**iss **M**e!

gives the order of the spectral classifications and is a useful aid.

A.1.4 Cosmology

Doppler effect

The lines in a star's absorption spectrum (page 82) are shifted when compared to the same lines as measured in a laboratory. This is due to the motion of the star relative to the Earth. If the star and the Earth are moving towards each other then the wavelengths of the absorption lines are shortened, i.e. shifted towards the blue end of the spectrum; we say that the wavelengths are **blue shifted**. Conversely, if the star and the Earth are moving away from each other then the wavelengths of the absorption lines are lengthened and moved towards the red end of the spectrum, or **red shifted**. This is known as the **Doppler effect**, and the relationship between the change in wavelength and the relative velocity of the object and the Earth is given by the Doppler formula:

$$\frac{\Delta\lambda}{\lambda} = \frac{\lambda_{app} - \lambda}{\lambda} = -\frac{v}{c}$$

where $\Delta\lambda$ is the change in wavelength, λ is the true wavelength of the absorption line, λ_{app} is the apparent wavelength of the absorption line on Earth, v is the relative velocity of the star and the Earth and c is the velocity of light. The relative velocity is taken to be positive when the two objects are approaching one another. The Doppler formula can also be expressed in terms of the change in frequency. Since $v = f\lambda$ we obtain:

$$\frac{\Delta f}{f} = \frac{v}{c}$$

Examiners' Notes

The minus sign is not needed here since approaching objects cause an increase in the frequency and receding objects a decrease in the frequency. Both expressions are based on the non-relativistic approach to the Doppler shift, i.e. $v \ll c$. The Doppler formula works for shifts at radio frequencies as well.

Example

The hydrogen absorption line from the star Vega is observed to have a wavelength of 656.255 nm (λ_{app}) compared to the same line in the laboratory of 656.285 nm. Determine the velocity of the star Vega and state whether it is approaching us or receding from us.

Answer

$\Delta\lambda = \lambda_{app} - \lambda = 656.255 - 656.285 = -0.030$ nm

$$\frac{\Delta\lambda}{\lambda} = -\frac{v}{c}$$

$$v = -\frac{c\Delta\lambda}{\lambda} = -\frac{(3\times10^8)\times(-0.030\times10^{-9})}{656.258\times10^{-9}} = 1.37\times10^4\,\text{m s}^{-1}$$

Since v is positive, Vega is approaching the Earth with a velocity of 13.7 km s^{-1}.

Doppler effect and the motion of binary stars

The Doppler effect can be used to determine the rotational velocity and the distance between two stars in a **binary star** system. Roughly half the stars found are in a binary system, in which two companion stars orbit their common centre of mass with periods ranging from hours to many thousands of years. The classification of binary stars is dependent upon the nature of the observation (most binaries are not resolved even with powerful telescopes); here we will consider spectroscopic binaries, as revealed by the Doppler shift of lines in their spectrum. To avoid further complications only **eclipsing binaries** will be looked at, i.e. those whose orbit lies in the same plane as the line of sight from Earth. These binary systems can be identified by their light curve (Fig 24). As one star eclipses the other, the apparent brightness of the combined binary image decreases.

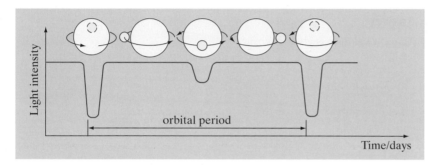

Fig 24
Light curve from an eclipsing binary system

Consider a binary system with one bright star and one faint star. The absorption lines will be Doppler shifted as the stars rotate about their centre of mass, moving between longer and shorter wavelengths in a periodic motion (Fig 25).

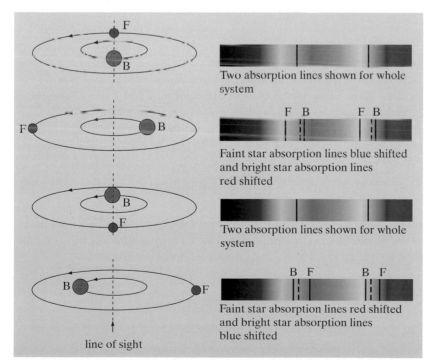

Fig 25
The sequence of changing positions of spectral lines as two stars rotate about each other in an eclipsing binary system. B is a bright star and F is a faint star. The amount of spectral shift depends on the rotational velocity

Essential Notes

Circular motion including angular velocity, frequency, and Newton's Law of Gravitation are covered in Unit 4.

Analysis of the spectral motion reveals a cyclic movement of a particular absorption line, shifted one way and then the next with a constant period. It is possible to calculate the angular velocity, period and distance between the two stars using the mechanics of circular motion combined with Newton's Law of Gravitation.

Example

Spectroscopic data on the Algol eclipsing binary star system show a period $T = 68$ hours for the K absorption line of calcium. In a laboratory the absorption line appears at 393.4 nm. The maximum periodic shift in the line from the star system is ± 0.3 nm. Calculate the angular velocity of the system, the linear (tangential) velocity and the distance between the stars.

Answer

Angular velocity is given by:

$$\omega = \frac{2\pi}{T} = \frac{2\pi}{68 \times 3600} = 2.56 \times 10^{-5}\,\text{s}^{-1}$$

Linear velocity is given by the shift in the spectral line:

$$\frac{\Delta\lambda}{\lambda} = -\frac{v}{c}$$

$$v = -\frac{c\Delta\lambda}{\lambda} = -\frac{(3 \times 10^8) \times (\pm 0.3 \times 10^{-9})}{393.4 \times 10^{-9}} = \pm 2.29 \times 10^5\,\text{m s}^{-1}$$

$v = r\omega$, so

$$r = \frac{v}{\omega} = \frac{2.29 \times 10^5}{2.56 \times 10^{-5}} = 8.94 \times 10^9\,\text{m}$$

Doppler effect and the recession of galaxies

Doppler shifts and wavelength changes can also be applied to whole galaxies. In the vast majority of cases the absorption (or emission) spectra from distant galaxies are found to be red shifted. This indicates that these galaxies are moving away from us and is evidence of an expanding Universe. In this case the **red shift** of a galaxy, denoted by the letter z, is given by:

$$z = \frac{\lambda_{\text{app}} - \lambda}{\lambda} = \frac{\Delta\lambda}{\lambda}$$

where λ_{app} is the observed (apparent) wavelength and λ is the wavelength of the spectral line as determined on Earth (the true wavelength). This can then be related to the recession velocity by the Doppler formula.

$$z = \frac{\text{recession velocity}}{\text{speed of light}} = -\frac{v}{c}$$

The above equation is only valid when the recession velocity is much less than the velocity of light, $< 0.1c$.

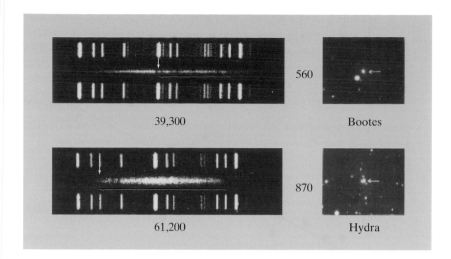

Fig 26
The optical spectra for two elliptical galaxies. Both have been taken with the same magnification. The yellow arrow indicates a pair of dark absorption lines that are shifted to longer wavelengths (red shifted). The figures on the right give the distance of the galaxy in Mpc and those below each spectrum give the recession velocity in km s^{-1}

Example

The K absorption line in singly ionised calcium normally has a wavelength of 393.4 nm. In spectra from galaxy NGC 4889 the line occurs at 401.8 nm. Determine the red shift of this galaxy and the recession velocity.

Answer

Here we have $\lambda = 393.4$ nm and $\lambda_{app} = 401.8$ nm, therefore

$$\Delta\lambda = \lambda_{app} - \lambda = 401.8 - 393.4 = 8.4 \text{ nm}$$

$$z = \frac{\Delta\lambda}{\lambda} = \frac{8.4}{393.4} = 0.0214$$

$$v = -cz = -(3 \times 10^5) \times 0.0214 = -6.42 \times 10^3 \text{ km s}^{-1}$$

The galaxy Hydra (see Fig 26) shows a recession velocity of $61\,000 \text{ km s}^{-1}$, which is $0.2c$ and above the threshold for an accurate determination of v; in this case an error in excess of 12% is introduced. To derive a more accurate value of v and hence z a relativistic formula has to be used.

Cosmologists do not think of galaxies moving through space away from us at such speeds, but regard space itself to be expanding and the light waves being stretched along with it. The relativistic version of the red shift is used to take account of this expansion and is called the **cosmological red shift**. Current measurements indicate cosmological red shifts as high as 10, which corresponds to a huge shift in wavelength, extremely high recession velocities and objects that lie at cosmological distances.

Hubble's Law

As discussed above, the spectra from galaxies (apart from a few very near to our own Milky Way) all show red shift. A plot of the recession velocity against distance for galaxies is close to a straight line (Fig 27) and is called a **Hubble diagram**, named after Edwin Hubble who discovered the relationship.

Essential Notes

Hubble's original data was based on the distances of Cepheid variables and therefore only extended to distances of ~2 Mpc. Current data has extended this to include galaxies as far distant as 1000 Mpc, where recession velocities are extremely high and the relativistic expression is used to determine v.

(2% of the total) have all been manufactured in the nuclear fusion processes that take place within stars and by supernova explosions that put these heavier elements back into the interstellar medium.

Quasars

Quasars, or 'quasi-stellar objects' (QSOs), show extremely large optical red shifts and so are believed to be some of the most distant objects in the known Universe. The earliest quasars discovered were strong radio emitters (hence the original name of QUASi-stellar Radio Sources) but this applies only to about a quarter of all quasars known today. Many emit their energy in the infra-red. More than 30 000 quasars have been detected, many with red shifts well in excess of $0.1c$ giving recession velocities in excess of $4 \times 10^7 \, \text{m s}^{-1}$ and distances of more than 700 Mpc. The farthest quasar detected is some 9000 Mpc away.

Optically these objects are very faint, but application of the inverse square law reveals them to be amongst the brightest objects in the Universe. Quasar 3C 273 has a luminosity of about 10^{40} W, comparable to 20 trillion Suns, and this is typical of many quasars. Overall quasar luminosities range from 10^{38} to 10^{42} W.

Other unusual properties of quasars include irregular variation in brightness from hours to weeks to months, and long narrow 'jets' of radiation. The elongated jets are thought to be caused by relativistic electrons spiralling in the local magnetic field and emitting bright beams of **synchrotron radiation** over all wavelengths.

Astronomers today regard quasars as intensely bright cores of distant **active galaxies**, perhaps with a supermassive black hole at the centre, but this question of their exact nature has yet to be answered. The answer has important consequences in respect of dark energy and dark matter, and hence for the case of an accelerating, expanding Universe.

Example

Quasar 3C 273 is known to have red shift of 0.16. Using a Hubble constant $H_0 = 73 \, \text{km s}^{-1} \, \text{Mpc}^{-1}$, determine how far away this quasar is from the Earth in Mpc.

Answer

Using the non-relativistic formula for the red shift gives a recession velocity of

$$v = cz = 3 \times 10^5 \times 0.16 \, \text{km s}^{-1}$$
$$= 4.8 \times 10^4 \, \text{km s}^{-1}$$

Using Hubble's Law then gives

$$d = \frac{v}{H_0} = \frac{4.8 \times 10^4}{73} = 658 \, \text{Mpc}$$

(In fact for red shifts >0.1 the relativistic equation for red shift should be used, but in this example the error introduced by using the non-relativistic form is $<10\%$.)

Practice exam-style questions

1 The Liverpool Telescope (LT) on the island of La Palma is the world's largest robotic telescope. It is a Cassegrain telescope with a primary mirror diameter of 2 m. It is used to collect electromagnetic radiation of wavelengths in the range 200 nm to 20 μm. All telescopes are limited in their resolving power, causing images to become blurred. A measure of this is given by the telescope's angular resolution.

(a) What is meant by the term *Rayleigh criterion* and how is this related to angular resolution?

_____

_____ 2 marks

(b) What is the *Airy disc*?

_____ 1 mark

(c) Calculate the maximum angular resolution of the LT.

_____

_____ 2 marks

(d) The LT is situated at a height of 2400 m above sea level. Which part of the electromagnetic spectrum is significantly absorbed by water vapour?

_____ 1 mark

Total Marks: 6

2 The ROSAT X-ray satellite detected an X-ray burst in the star cluster Terzan 2 with an apparent luminosity of 10^{30} W measured at a peak wavelength of 0.3 nm.

(a) Using Stefan's Law, give a formula for the radius of a star, R, in terms of its temperature and luminosity, assuming the star is a sphere.

_____

_____ 3 marks

(b) Calculate the temperature, T, of this star using Wien's Displacement Law.

_____

_____ 2 marks

(c) Use this temperature in your formula from part (a) to determine an approximate radius of this star. ($\sigma = 5.7 \times 10^{-8}$ W m^{-2} K^{-4})

_____

_____ 2 marks

(d) What type of star could Terzan 2 be?

_____ 1 mark

Total Marks: 8

3 Distant galaxies are observed to show red shifts, and the farther away a galaxy is the greater the red shift that is observed.

(a) State what is meant by the term *red shift*.

_____

_____ 1 mark

(b) How is the red shift of a galaxy determined?

_____

_____

_____ 2 marks

(c) Hubble's Law states that the velocity of recession of a galaxy is proportional to the distance to the galaxy as measured from the Earth. The table shows distances and recession velocities for three galaxies. Using Hubble's Law estimate the value of the Hubble constant H_0.

Distance Mpc	Velocity of recession $km\,s^{-1}$	H_0 $km\,s^{-1}\,Mpc^{-1}$
70	4000	
110	7500	
35	2200	

Average value of H_0 = _____ 4 marks

(d) The value of $1/H_0$ gives an estimate of the time passed since all the galaxies were close together; this gives an estimate for the age of the Universe. Use your average value of H_0 to estimate the age of the Universe. (1 year $= 3.2 \times 10^7\,s$, 1 pc $= 3.09 \times 10^{16}\,m$)

_____

_____

_____ 3 marks

(e) Suggest one reason why the age of the Universe may be greater than this value.

_____

_____

_____ 1 mark

Total Marks: 11

4 The star Betelgeuse in the constellation Orion has an apparent magnitude of +0.45 and an absolute magnitude of −5.1.

(a) What is meant by the terms *apparent magnitude* and *absolute magnitude*?

_____

_____ 2 marks

(b) Determine the distance to the star Betelgeuse in Mpc.

_____

_____

_____ 3 marks

(c) How much more luminous is Betelgeuse than the Sun? (The absolute magnitude of the Sun is +4.83.)

_____

_____

_____

_____ 4 marks

(d) The diagram shows the axes of a Hertzsprung-Russell (H-R) diagram

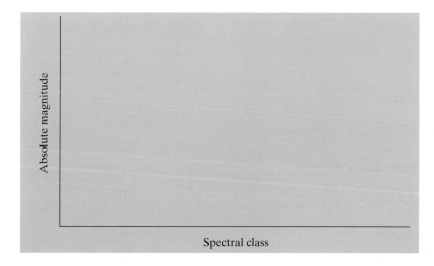

(i) On each axis indicate a suitable range of values.

(ii) Label the position of the main sequence, dwarf stars and giant stars.

(iii) Label the position of the Sun with the letter S.

(iv) Label the position of Betelgeuse with the letter B.   4 marks

Total Marks: 13

Answers, explanations, hints and tips

Question	Answer		Marks
1 (a)	The critical condition in which two sources can just be resolved and given by $\theta \approx \dfrac{\lambda}{D}$ where θ is the angular resolution.	(1)	
	This is met when the first minimum of the diffraction pattern of one source coincides with the centre of the diffraction pattern of the other.	(1)	2
1 (b)	This is the bright central maximum of the diffraction pattern of a point source.	(1)	1
1 (c)	Using		
	$\theta = \dfrac{\lambda}{D}$ for $\lambda = 200\,\text{nm}$,	(1)	
	$\theta = \dfrac{200 \times 10^{-9}}{2} = 1.0 \times 10^{-7}$ rad	(1)	2
1 (d)	Infra-red or microwaves.	(1)	1
			Total 6
2 (a)	Stefan's Law: $L = \sigma A T^4$		
	$\therefore L = \sigma(4\pi R^2) T^4$	(1)	
	$R^2 = \dfrac{L}{4\pi\sigma T^4}$	(1)	
	$R = \left(\dfrac{L}{4\pi\sigma T^4}\right)^{\frac{1}{2}}$	(1)	3
2 (b)	Wien's Displacement Law: $\lambda_{max} T = 2.9 \times 10^{-3}\,\text{m K}$		
	$\therefore T = \dfrac{2.9 \times 10^{-3}}{0.3 \times 10^{-9}}$	(1)	
	$= 9.7 \times 10^6\,\text{K}$	(1)	2
2 (c)	$R = \left(\dfrac{L}{4\pi\sigma T^4}\right)^{\frac{1}{2}} = \left(\dfrac{10^{30}}{4\pi \times 5.7 \times 10^{-8} \times (9.7 \times 10^6)^4}\right)^{\frac{1}{2}}$	(1)	
	$R = 1.26 \times 10^4\,\text{m}$	(1)	2
2 (d)	A neutron star.	(1)	1
			Total 8

Question	Answer		Marks
3 (a)	The increase in wavelength of electromagnetic radiation emitted by an object that is moving away from the observer; wavelength appears longer hence redder.	(1)	1
3(b)	It is determined from the shift in the absorption lines,	(1)	
	and using the Doppler formula:		2
	red shift $z = \dfrac{v}{c} = \dfrac{\Delta\lambda}{\lambda}$	(1)	
3(c)	Hubble's Law gives $H_0 = v/d$.		

Distance Mpc	Velocity of recession $km\,s^{-1}$	H_0 $km\,s^{-1}\,Mpc^{-1}$
70	4000	57.1
110	7500	68.2
35	2200	62.9

		(3)	
	Average value of $H_0 = 62.7\ km\,s^{-1}Mpc^{-1}$	(1)	4
3(d)	Change the units of H_0 into s^{-1}.		
	$H_0 = 62.7\ km\,s^{-1}\,Mpc^{-1} = \dfrac{62.7 \times 10^3}{3.09 \times 10^{16} \times 10^6}$	(1)	
	$H_0 = 2.03 \times 10^{-18}\ s^{-1}$	(1)	
	Estimated age of Universe $= \dfrac{1}{H_0} = 4.9 \times 10^{17}\ s$		3
	$= 1.54 \times 10^{10}$ years $= 15.4$ billion years	(1)	
3(e)	The above calculation is based on the assumption that the expansion has been constant throughout this time. In fact the expansion is accelerating, so it will have been slower in the past, making the age greater than calculated.	(1)	1
			Total 11

Question	Answer		Marks
4(a)	The apparent magnitude is the apparent brightness of a star expressed by the magnitude scale.	(1)	
	The absolute magnitude is the apparent magnitude a star would have if it were placed at a standard distance of 10 parsec.	(1)	2
4(b)	Using $$m - M = 5\log\frac{d}{10}$$		
	$$0.45 - (-5.1) = 5\log\frac{d}{10}$$	(1)	
	$$\therefore \log\frac{d}{10} = \frac{5.55}{5} = 1.11, \text{ and } \frac{d}{10} = 10^{1.11}$$	(1)	
	$d = 129$ Mpc	(1)	3
4(c)	$$M - M_\odot = -2.5\log\left(\frac{L}{L_\odot}\right)$$	(1)	
	$$-5.1 - 4.83 = -2.5\log\left(\frac{L}{L_\odot}\right)$$	(1)	
	$$\log\left(\frac{L}{L_\odot}\right) = 3.972$$	(1)	
	$$L = L_\odot \times 10^{3.972} = 9376 L_\odot$$	(1)	
	i.e. Betelgeuse is about 9400 times more luminous than the Sun.		4
4(d)	Show the areas of each region clearly.		4
			Total 13

Option Unit 5B Medical Physics

B.2.1 Physics of the eye

The human eye is a remarkable optical system, a living video camera. It contains 130 million light-detecting cells. These are sensitive enough to form an image in the darkness of night and adaptable enough to cope with summer sunshine, when the light may be 10 million times brighter. With perfect eyesight, you can see details as small as 1 mm up to 6 m away and switch your focus from a 20 cm close-up to infinity in less than one-tenth of a second.

This section deals with the eye as an optical system, explains how it forms an image and discusses some of the common defects of vision and how they can be corrected.

Physics of vision

The human eye and a video camera (Fig 1) have similar demands. They need to:

- produce an image in 'real time', which means updating the image around 20 times per second

- automatically adjust the focus from close-up to infinity

- automatically adapt to changing light conditions.

There are similarities about the way that the eye and the camera meet these demands. They both admit light through a variable aperture, and they both use a convex lens to focus an image onto a light-sensitive surface.

Fig 1
(a) The eye–brain system compared with (b) the video camera–TV system

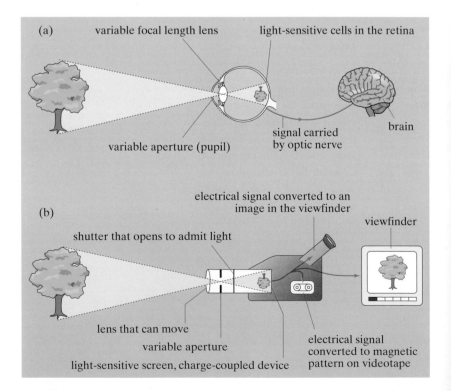

Controlling the brightness

The amount of light entering a video camera is controlled by a variable aperture between the lens and the charge-coupled device (CCD) that detects the light. In the eye (Fig 2), the aperture is in front of the lens and is known as the **pupil**. The diameter of the pupil is varied by a ring of smooth muscle known as the **iris**. The iris is the coloured part of your eye. Some of the muscle fibres in the iris run radially, like bicycle spokes. Other muscle fibres encircle the pupil. When the circular fibres contract, the pupil becomes smaller (constricts). When the radial fibres contract, the pupil opens wider (dilates).

Essential Notes

The iris is the coloured part of your eye. The colour comes from a pigment that absorbs light. People with darker eyes just have more pigment in their iris.

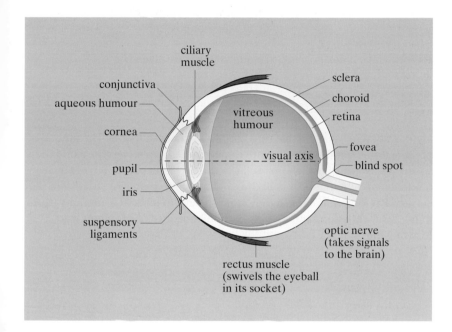

Fig 2
Vertical section through the eye

Example

The eye can increase its sensitivity to light by a factor of about 1 million. The diameter of the pupil can vary from about 1.5 mm in bright light to about 8.0 mm in low-light conditions. Is the increase in sensitivity entirely due to the change in area?

Answer

Area when constricted $= \pi r^2 = \pi(0.75 \times 10^{-3})^2$
Area when dilated $= \pi r^2 = \pi(4.0 \times 10^{-3})^2$
The ratio of these is

$$\frac{\pi(4.0 \times 10^{-3})^2}{\pi(0.75 \times 10^{-3})^2} = \left(\frac{4.0}{0.75}\right)^2 = 28$$

The area changes by a factor of 28, so there must be another, much more important, mechanism for increasing sensitivity.

Essential Notes

Refraction of light was covered in Unit 2.

Fig 3
Refraction at a boundary. The angle of refraction, r, of the light depends on the angle of incidence, i, and on the refractive index of the medium

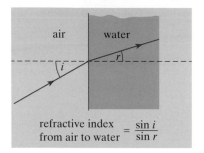

$$\text{refractive index from air to water} = \frac{\sin i}{\sin r}$$

Forming a focused image

Light rays enter the eye through the **cornea**, the transparent membrane that covers the front of the eye (Fig 2). If the cornea gets scratched, it can repair itself. But it can become opaque as a result of illness or accident. In these cases, sight may be restored by a cornea transplant.

Rays of light are deviated when they cross a boundary between two materials (Fig 3) because of refraction, which is due to the changing speed of light in the different media. When a ray travels from medium 1 to medium 2, the **refractive index**, $_1n_2$, is defined as

$$_1n_2 = \frac{\text{speed of light in medium 1}}{\text{speed of light in medium 2}}$$

The refractive index from air to water is therefore given by:

$$_{\text{air}}n_{\text{water}} = \frac{3.00 \times 10^8 \text{ m s}^{-1}}{2.24 \times 10^8 \text{ m s}^{-1}} = 1.34$$

Light is refracted as it passes from air into the cornea (Fig 4). The amount of refraction depends on the relative speed of light in air as compared to its speed in the cornea, which is similar to that in water. About 60% of the refraction of light that occurs in the eye happens when light strikes the cornea. The cornea bulges slightly from the eye to present a curved surface to the light. Variations and defects in this shape lead to problems in focusing the light.

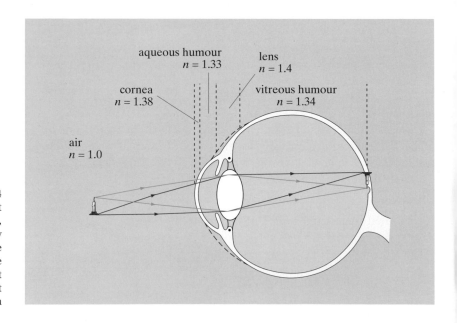

Fig 4
Refraction in the eye. As a ray of light passes through each part of the eye, it is refracted at each boundary between media. A large change in the refractive index causes more refraction. In the eye, the largest deviation will occur as the light passes from air into the cornea

After the cornea, light passes through a clear watery fluid known as the aqueous humour before reaching the lens. The pressure in this fluid is normally about 15 mmHg (2000 Pa) above atmospheric pressure, which helps the eyeball to maintain its shape.

Lenses

The lens in the eye is a **converging lens**—it refracts parallel rays of light so that they come together. The principal focus of a converging lens is the point at which parallel rays of light, at right angles to the lens, are brought to a focal point. The focal length is the distance from the centre of the lens to the principal focus. The focal length of a lens is determined partly by its shape: a lens that is highly curved (small radius of curvature) will have a short focal length.

Essential Notes

The lens in the eye is **convex**. This refers to its shape—it is thicker in the middle than at the edges. Convex lenses converge rays of light. Lenses that are thinner in the middle than at the edges are known as **concave**. Concave lenses diverge (spread apart) light rays.

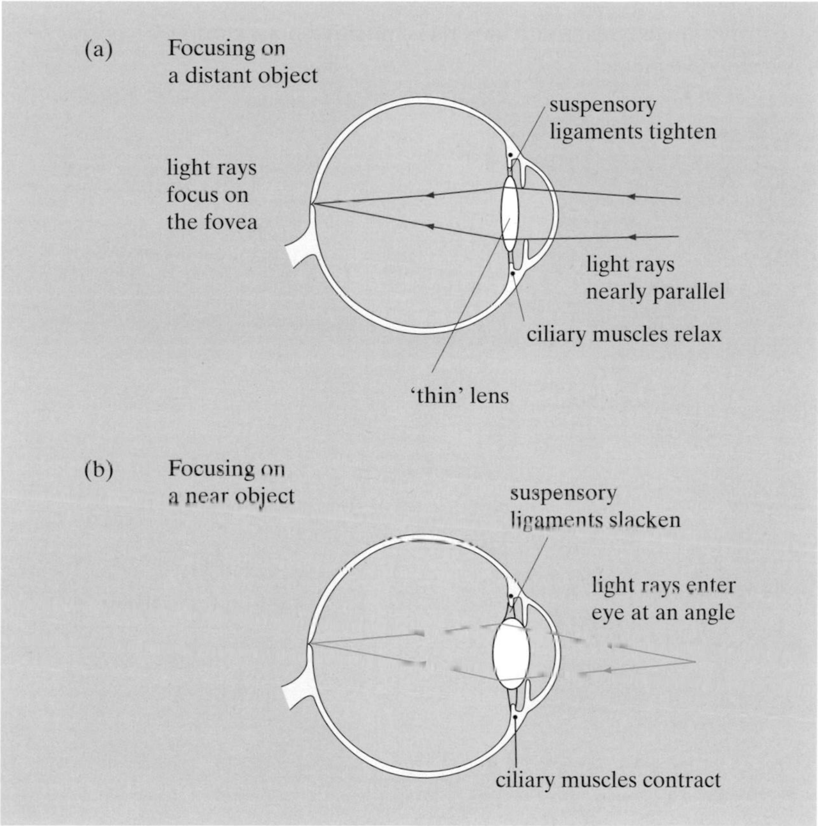

(a) Focusing on a distant object

suspensory ligaments tighten

light rays focus on the fovea

light rays nearly parallel

ciliary muscles relax

'thin' lens

(b) Focusing on a near object

suspensory ligaments slacken

light rays enter eye at an angle

ciliary muscles contract

Fig 5
Focusing in the eye for an object that is (a) distant and (b) near

Essential Notes

The lens of our eye is formed entirely of transparent cells, each measuring about 1 cm in length but with a thickness of only 2 μm.
As we age, the lens becomes less flexible and loses its ability to focus at different distances.

In the eye, the lens brings rays of light from an object to a focus at the **fovea** on the **retina** at the back of the eye (Fig 5a). The lens makes the small adjustments to the direction of the light rays that are needed to produce this focused image. The lens is held in tension by ligaments, which are connected to the ciliary muscle. As this muscle contracts and then relaxes, the lens changes shape (Fig 5b) and therefore changes its focal length.

The ability of the lens in the eye to change its focal length, so that objects at different distances are brought into sharp focus on the retina, is called **accommodation**. When the lens adopts a highly curved surface, it increases its focusing power. The **power** of a refracting surface is measured in **dioptres** (D), and is defined as

$$\text{power (dioptres)} = \frac{1}{\text{focal length (metres)}} = \frac{1}{f}$$

By convention, a converging lens has a positive power, and a diverging lens has a negative power (Fig 6).

Fig 6
The power of converging and diverging lenses

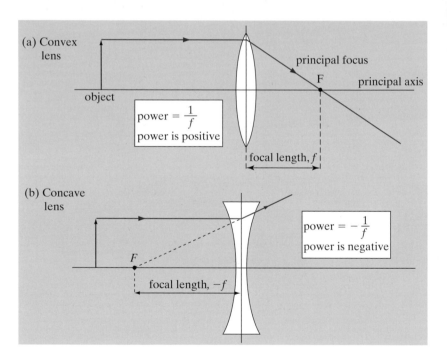

The total refracting power of a combination of lenses, or surfaces, is the sum of their powers. The human cornea has a refracting power of about $+43\,\text{D}$. The power of the lens varies from $+17\,\text{D}$ at its flattest to about $+31\,\text{D}$ at its most curved. The total power of the eye therefore varies from about $+60\,\text{D}$ to $+74\,\text{D}$.

There is a limit on the range of distances at which the eye can focus. The **near point** is the closest distance at which an object can be brought into focus. For a young, healthy eye, the near point is at about 25 cm. For older people, the increasing stiffness of the lens makes accommodation more difficult, and the near point becomes further away. This loss of accommodation is known as **presbyopia**. The **far point** is the furthest distance at which an object can be brought into focus. For a normal eye, the far point is infinity.

Ray diagrams

For a lens of known power, it is possible to find out where an image will be formed by constructing a ray diagram. Figs 7 and 8 show ray diagrams for a convex lens and a concave lens, respectively. In a ray diagram, the lens can be represented as a straight vertical line. A horizontal line drawn through the centre of the lens and at right angles to it is called the **principal axis**. The object is represented as an arrow and the diagram is drawn to scale. The principal focus of the lens is marked on the principal axis on both sides of the lens. This is shown with the symbol F in Figs 7 and 8. Rays of light leave the tip of the object in every direction. For three specific rays drawn leaving the object, we can predict their passage through the lens and locate the focused image.

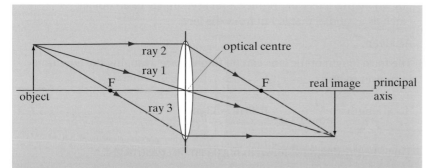

Ray 1   Any ray of light that passes through the optical centre will be undeviated

Ray 2   A ray of light that travels parallel to the principal axis will be refracted through the principal focus

Ray 3   Any ray of light that passes through the principal focus will be refracted so as to travel parallel to the principal axis (this is rule 2 in reverse)

Fig 7
Ray diagram of light passing through a convex lens. The focused image is formed where the rays of light from the tip of the object meet again. This is referred to as a **real image**, because rays of light actually pass through it – an image would be formed on a screen placed at this point

Essential Notes

For ease of drawing, the rays in such diagrams are shown as bending only at the centre line, though in reality there will be refraction at both interfaces. Also the lens is often drawn as just a vertical line.

It is possible for an image distance to be negative. If the object in the example above was placed 5.0 cm from the lens, the calculation would become:

$$\frac{1}{v} = \frac{1}{f} - \frac{1}{u} = \frac{1}{0.100} - \frac{1}{0.050} = 10.0 - 20.0 = -10.0\,\text{m}^{-1}$$

So

$$v = \frac{1}{-10} = -0.10\,\text{m}$$

The image distance is negative. In this case the lens is not powerful enough to bring the image to a focus at all. If we looked through this lens towards the object, the light rays would appear to come from a point $-0.10\,\text{m}$ from the lens. A virtual image has been formed (Fig 11). No light rays actually pass through this image, and it could not be formed on a screen. A negative value of the image distance, v, means that a virtual image has been formed on the *same* side of the lens as the object.

Fig 11
Formation of a virtual image by a convex lens

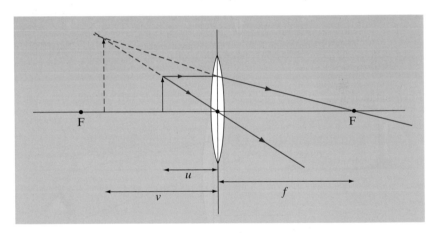

Examiners' Notes

Follow the 'real is positive' convention. A positive number for image distance means a real image, whereas a negative number means a virtual image. A converging lens has a positive focal length, whereas a diverging lens has a negative focal length.

The lens formula also works for diverging lenses. However, we use a negative number to represent the focal length of a diverging lens.

Magnification

The image may be smaller or larger than the object. The **magnification** of an image produced by a lens is the ratio of image height to object height. So, in Fig 12, we have

$$\text{magnification} = \frac{XY}{AB}$$

But

$$\tan AOB = \frac{AB}{AO} \quad \text{and} \quad \tan XOY = \frac{XY}{XO}$$

Angles AOB and XOY are equal, so

$$\frac{AB}{AO} = \frac{XY}{XO}$$

This means that

$$\frac{XY}{AB} = \frac{XO}{AO}$$

and this gives an alternative definition for magnification.

Definition

Magnification m is given by

$$m = \frac{\text{image distance}}{\text{object distance}} = \frac{v}{u}$$

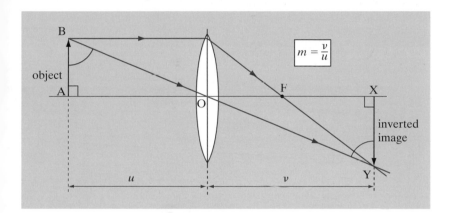

Fig 12
The definition of magnification

Example

A convex lens with a focal length of 20 cm is to be used to produce an image of an object that is 30 cm high and 1 m away from the lens. How big will the image be?

Answer

Use the lens formula to find the image distance, v. Rearranging the lens formula gives

$$\frac{1}{v} = \frac{1}{f} - \frac{1}{u} = \frac{1}{0.20} - \frac{1}{1.0} = 5.0 - 1.0 = 4 \text{ m}^{-1}$$

so

$$v = \frac{1}{4} = 0.25 \text{ m} = 25 \text{ cm}$$

The magnification m is

$$m = \frac{v}{u} = \frac{25}{100} = 0.25$$

so the image will be $0.25 \times 30 = 7.5$ cm high.

Defects of vision and their correction using lenses

Short and long sight

Short-sighted people can focus on nearby objects (Fig 13a), but cannot focus on distant ones (Fig 13b). They are said to have **myopia**. Someone suffering from myopia has a cornea and lens combination that is too powerful, or an eyeball that is too long. Distant objects are brought to a focus before the light reaches the retina (Fig 13b). Short-sighted people need to use a diverging lens to correct their vision (Fig 13c).

Fig 13
The short-sighted or myopic eye

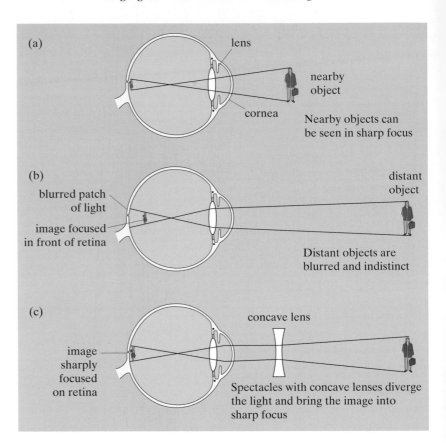

Example

Sarah is short-sighted. Without glasses, her far point is 50 cm. If the distance between her lens and retina is 20 mm, find the power of the lens needed to correct her vision.

Answer

Using the lens formula, we can work out the power of Sarah's eye when she tries to focus on a distant object. The image distance is $v = 0.02$ m and the object distance is $u = 0.50$ m, so

$$\text{power} = \frac{1}{f} = \frac{1}{0.50} + \frac{1}{0.02} = 52 \, \text{D}$$

If Sarah is to focus on an object at infinity, then $u = \infty$, and the power needs to be

$$\text{power} = \frac{1}{f} = \frac{1}{\infty} + \frac{1}{0.02} = 50\,\text{D}$$

We need to reduce the total refracting power of her eyes by 2 D, so she needs lenses of power $-2\,\text{D}$.

When the lens–cornea combination is not powerful enough, the light from nearby objects cannot be brought to a focus on the retina (Fig 14b). This is long sight, or **hypermetropia**. Someone with hypermetropia can focus on objects at infinity (Fig 14a), but not on nearby things (Fig 14b). A convex lens is needed to converge the light more (Fig 14c).

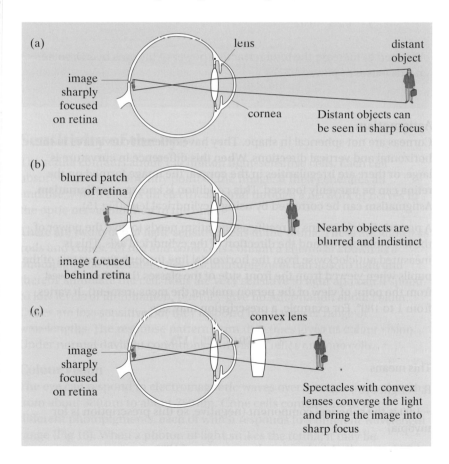

(a)

lens

distant object

image sharply focused on retina

cornea

Distant objects can be seen in sharp focus

(b)

blurred patch of retina

image focused behind retina

Nearby objects are blurred and indistinct

(c)

convex lens

image sharply focused on retina

Spectacles with convex lenses converge the light and bring the image into sharp focus

Fig 14
The long-sighted or hypermetropic eye

B.2.2 Physics of the ear

The ear as a sound detection system

Sound waves are longitudinal vibrations that travel through a medium, such as air, as a pressure wave (Fig 20). The energy transferred by the wave is proportional to the square of its amplitude.

Fig 20
Sound waves

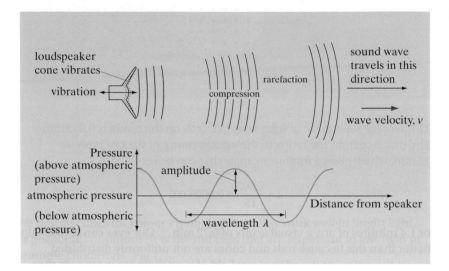

The human ear is an incredibly sensitive device for collecting and detecting sound waves.

The visible part of the human ear is known as the pinna (Fig 21). Its function is to collect sound waves and funnel them down the external auditory tube. The pinna is shaped so that sound sources in front of the head are detected more easily than those behind. This helps us to determine the direction of a sound source.

- The *outer ear* collects sound waves and relays them to the eardrum.

- The *middle ear* transmits the vibrations of the eardrum to the inner ear, amplifying or damping the vibrations as necessary.

- The *inner ear* or **cochlea** converts the vibrations to electrical signals, which are transmitted to the brain via the auditory nerve.

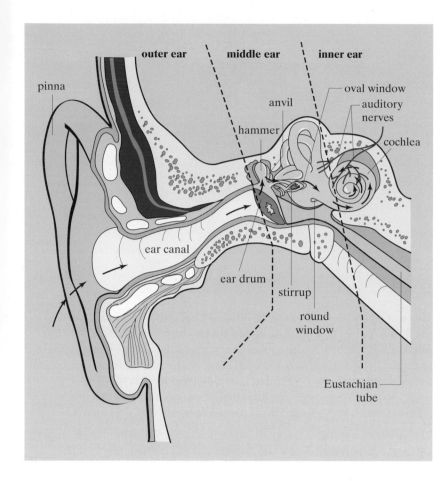

Fig 21
A cross-section through the ear.
A sound wave passes through the
three sections of the ear in about
20 ms (20 milliseconds)

Essential Notes

One cause of hearing loss is an
obstruction in the ear canal.
Foreign bodies, or too much
wax, are common problems.

Fig 22
When the air in the external auditory
tube vibrates in its fundamental
mode (resonates), the wavelength
of the sound is $\frac{\lambda}{4} = l = 2.5\,\text{cm}$,
so $\lambda = 10\,\text{cm}$.
The speed of sound in air is about
$v = 330\,\text{m s}^{-1}$. Since $v = f\lambda$, then

$$f = \frac{v}{\lambda} = \frac{330}{10 \times 10^{-2}} = 3300\,\text{Hz}$$

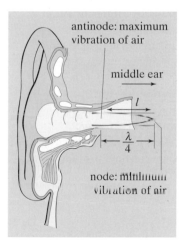

The external auditory tube, or ear canal, is about 2.5 cm long and has a
diameter of 7 mm. The ear canal acts like a miniature organ pipe, and
modifies the sound we hear through its resonant properties (Fig 22). The
ear has its maximum response at around a frequency of 3300 Hz.

The sound waves in the ear canal cause the eardrum to vibrate. These
vibrations have a remarkably small amplitude. For very quiet sounds, your
eardrum may vibrate by only 10^{-11} m (less than the diameter of an atom).

The bones of the middle ear (ossicles) are the malleus (hammer), incus
(anvil) and stapes (stirrup), as seen in Fig 23. They transfer the oscillations
from the eardrum to the oval window, which is the entrance to the cochlea.

Essential Notes

The eardrum can be damaged
by very loud sounds, such as a
rifle being fired close to the ear.
Sudden pressure changes,
perhaps caused by a blow to
the ear, can also cause damage.

Fig 23
The middle ear. The bones of the middle ear transmit the vibrations of the eardrum to the oval window

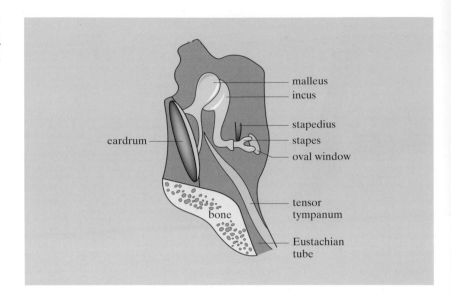

Essential Notes

The small muscles that are attached to the stapes and malleus can protect our ears from loud noises to some extent. When the muscles contract, they reduce the vibrations passed to the inner ear. The process takes about 50 ms and so cannot protect against sudden changes in sound volume.

Essential Notes

The Eustachian tube connects the middle ear to the throat. It is normally closed, but opens when we swallow or yawn. The tube ensures that the air pressure is the same as in the outer ear. Any pressure differences tend to reduce sensitivity and cause pain. It is fairly common for the tube to become blocked in children, and the middle ear then fills with fluid. Small tubes (grommets) are inserted to allow the fluid to drain away.

These bones enable vibrations in the air of the outer ear to be passed to the liquid in the inner ear. The bones act as a lever, which magnifies the force by a factor of 1.3. The area of the oval window is only one-twentieth of that of the eardrum, so the overall effect of the middle ear is to increase the pressure acting on the fluid in the middle ear by a factor of 26 ($= 1.3 \times 20$). Without this amplification, there would be significant acoustic losses as the sound wave passed between different media.

The vibrations of the oval window are transferred to the fluid in the inner ear as pressure waves, which are detected in the cochlea (Fig 24). The cochlea is a cavity, about the size of a pea, in the thickest part of the skull. It contains hair cells that convert vibrations to electrical impulses. There are about 30 000 sensory hair cells in each ear. They are amazingly sensitive: a distortion of 0.0003° can be detected.

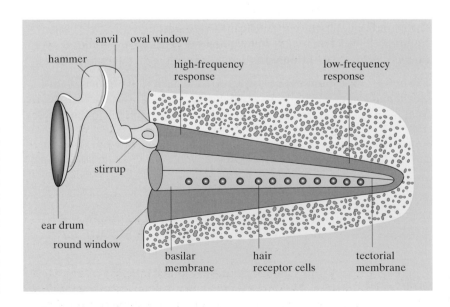

Fig 24
In this simplified diagram the cochlea has been unrolled and magnified. The cochlea contains two membranes, the basilar and the tectorial. The basilar membrane holds a row of hair cells. These hairs project through the fluid of the inner ear to the tectorial membrane. When the stapes vibrates against the oval window, the basilar membrane vibrates and the hair cells are distorted

Hair cells can fire quickly enough to detect sound waves up to a frequency of 1 kHz. Nerve cells cannot carry impulses faster than about 1 kHz. Another mechanism is needed to detect high-frequency sounds. Low-frequency sounds cause the whole length of the basilar membrane to vibrate, whereas high-frequency sounds cause just the first part of the membrane to vibrate (Fig 24). Different frequencies cause different sections of the hair cells to vibrate, which the brain interprets as pitch, signalling a high or a low note.

Sensitivity and frequency response

The ear's ability to detect small changes in intensity is known as its sensitivity. Our ears respond differently to different frequencies of sound, so the sensitivity is affected by frequency. Our ears are most sensitive at around 3 kHz. At this frequency, the ear can detect changes of intensity of around 12%.

The perceived loudness of a sound is a difficult thing to measure, because it depends on the listener as well as the sound. What is loud for one person may be barely audible for someone else. The sensation of loudness also depends on frequency. However, for a given person, the loudness of a sound at a certain frequency depends on the rate at which energy is transferred through a given area by the sound wave. This is known as the sound **intensity** and it is an objectively measurable quantity.

> **Definition**
>
> Sound **intensity** is the power (energy per second) through unit area perpendicular to the direction of propagation (travel) of the wave. It is measured in watts per square metre ($W\,m^{-2}$).

The intensity of sound from a point source follows the **inverse square law** (Fig 25). Sound spreads out from a point source as a spherical wave. The intensity of the wave is its power per unit area. As the wave travels further from the source, the area of the wavefront increases, and the intensity decreases. As the distance doubles from a sound source, the intensity drops to one-quarter $\left(\frac{1}{2}\right)^2$ of its initial value. If the distance is trebled, the intensity drops to one-ninth $\left(\frac{1}{3}\right)^2$

Fig 25
Sound and the inverse square law

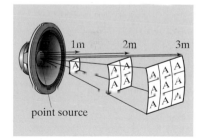

point source

The threshold of hearing

The human ear can detect and respond to an enormous range of intensities. At a frequency of 1 kHz, the lowest intensity that a normal human ear can detect is about $1 \times 10^{-12}\,W\,m^{-2}$. This intensity is defined as the **threshold of hearing**, I_0, so

$$I_0 = 1 \times 10^{-12}\,W\,m^{-2}$$

The loudest sound that we can detect without severe discomfort corresponds to an intensity of about $100\,W\,m^{-2}$. This is known as the 'threshold of pain' (see Fig 26, page 126).

Between these very quiet and very loud sounds, the intensity changes by a factor of 10^{14}. However, the sensation of loudness that we experience does not vary by this much. Our ears have an 'automatic volume control' that turns down the amplification when we are exposed to high-intensity sounds. The ear is not a linear device, so turning up the intensity of a sound wave in equal steps does not increase the loudness by equal amounts. When it is quiet, a small increase in the sound intensity, such as someone coughing in a library, would seem loud. The same increase in sound intensity at a rock concert would go unnoticed.

In fact, the change in the loudness that we hear is proportional to the fractional change in intensity:

$$\text{perceived change in loudness} \propto \frac{\text{change in intensity}}{\text{initial intensity}}$$

We hear the same change in loudness when the sound intensity increases from $1 \times 10^{-12}\,\mathrm{W\,m^{-2}}$ to $2 \times 10^{-12}\,\mathrm{W\,m^{-2}}$ as we do when it increases from $1\,\mathrm{W\,m^{-2}}$ to $2\,\mathrm{W\,m^{-2}}$. This means that the loudness L that we experience is proportional to the logarithm of the intensity, rather than to the intensity itself:

$$L \propto \log_{10}\left(\frac{I}{I_0}\right)$$

This logarithmic response gives the ear its enormous **dynamic range**. A healthy human ear can respond to noises as loud as a jet taking off and detect sounds as quiet as a pin dropping, a ratio of intensities of about 10^{14}.

Relative intensity levels of sounds

Intensity level and the decibel

Because the ear has a logarithmic response to intensity, we use a logarithmic scale to measure sound intensity. Sound intensity level is measured in **decibels** (dB) on the **dB scale**:

$$\text{sound intensity level (dB)} = 10\log_{10}\left(\frac{I}{I_0}\right)$$

where I_0 is the threshold of hearing. All sound levels are compared to the threshold of hearing. A sound that has twice the intensity of the threshold of hearing will have a sound intensity level given by

$$\text{sound intensity level (in dB)} = 10\log_{10}\left(\frac{2 \times 10^{-2}}{1 \times 10^{-12}}\right) = 10\log_{10} 2 = 3\,\mathrm{dB}$$

Doubling the sound intensity always corresponds to a change of 3 dB in the sound intensity level. By a similar calculation, adding 10 dB to the sound intensity level, means a factor of $10 \times$ the sound intensity.

The db scale measures sound intensity level, *not* loudness. A sound intensity level of 20 dB would seem louder to some people than to others.

Table 1
The dB scale is used to quantify everyday sounds

dB	Typical example
130–120	jet aircraft taking off
110	pneumatic drill 1 m away
100	disco
90	symphony orchestra at a crescendo
80	vacuum cleaner 1 m away
70	inside a moving bus
60	general classroom noise
50	whisper 1 m away
40	quiet classroom
30–10	very quiet countryside
0	threshold of hearing

Example

At a rock concert, the sound intensity level is 120 dB when you are 1 m from the speakers.

(a) Calculate the sound intensity at this distance.

(b) How far away would you have to be for the sound to drop to the 'safe' level of 90 dB?

Answer

(a) Use

$$\text{sound intensity level (dB)} = 10 \log_{10}\left(\frac{I}{I_0}\right)$$

So

$$\text{inverse } \log_{10}\left(\frac{\text{sound intensity level (dB)}}{10}\right) = \frac{I}{I_0}$$

$$10^{\left(\frac{120}{10}\right)} = \frac{I}{I_0}$$

$$I = I_0 \times 10^{12} = 1 \times 10^{-12} \times 1 \times 10^{12} = 1\,\text{W m}^{-2}$$

(b) The sound intensity that corresponds to 90 dB is $0.001\,\text{W m}^{-2}$, because it is 30 dB ($= 10^{-3}$) less than 120 dB (see answer to part (a)). The ratio of intensities is $0.001/1 = 0.001$. If we apply the inverse square law, this is a ratio of the distances $(r_1/r_2)^2$, and since $r_1 = 1$ m, we find

$$\left(\frac{1}{r_2}\right)^2 = 0.001$$

$$r_2^2 = \frac{1}{0.001} = 1000$$

$$r_2 = 31.6\,\text{m}$$

This assumes that the speakers behave as point sources of sound from which the sound waves spread out equally in all directions. In practice, this will not be the case. Loudspeakers are designed to focus the sound in front, rather than behind.

Essential Notes

Remember that inverse $\log_{10} x$ is the same as 10^x.

Equal-loudness curves

The actual range of frequencies that can be heard varies from person to person, but a young adult with healthy ears can typically detect sounds from 20 Hz to 20 kHz (Fig 26). Sounds above 20 kHz are referred to as ultrasonic or ultrasound. The threshold of hearing is defined as 0 dB at 1 kHz, but our ears can detect 3 kHz sounds at even lower intensities than this. Very high-frequency sounds can be detected if the intensity is high enough. Fig 26 shows how the minimum intensity of sound that can be heard varies with frequency. A sound at around 2–3 kHz will seem louder than one at 10 kHz, even though the intensities are the same.

Fig 26
The frequency range of the normal ear. The solid line represents the intensity level necessary for a sound at that frequency to be heard by a normal ear

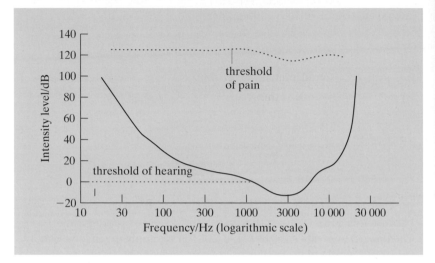

The loudness of a sound for a given person is measured by comparison with the loudness of a standard source of sound. A 1 kHz standard sound source is placed next to the source of unknown loudness. The intensity of the 1 kHz sound is adjusted until it sounds just as loud as the unknown source (Fig 27). If the 1 kHz source is then measured to be 70 dB, the unknown source is said to have a loudness of 70 phons.

Fig 27
These equal-loudness curves for the normal ear show the sound intensity level that is required to produce the same perception of loudness at different frequencies

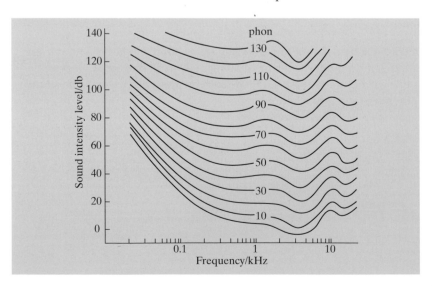

It is possible to take into account the frequency dependence of our hearing when we are trying to specify sound intensity levels. Sound level meters can have their output weighted so that they simulate the response of the human ear. The weighted sound intensity levels are a measure of the effect on the ear and are represented by the **dBA scale** (Fig 28). The dBA scale is used for environmental monitoring.

Defects of hearing

The sensitivity of our hearing deteriorates as we get older. Hearing may also be damaged by exposure to loud noises. Hearing damage can be assessed by testing a person's response to sounds at different frequencies. An **audiometer** is used to produce sounds at various frequencies and intensities. These are played to the person under test through headphones, or using an electronic vibrator held on the bone behind the ear. The audiometer is adjusted to read 0 dB at the intensity level at which the sound can just be detected by a person with normal hearing. If a patient can only detect sounds at this frequency that are 50 dB louder than this, they are said to have a hearing loss of 50 dB. This test is carried out at a number of frequencies. The resulting plot of frequency against hearing loss is known as an **audiogram**. Some typical audiograms are shown in Fig 29.

Fig 28
The dBA scale. These weightings are used to simulate the response of the human ear. The sound level meter is therefore less sensitive to low and very high frequencies

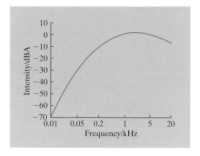

Essential Notes

If the use of the vibrating device on the bone shows a normal audiogram, but the headphones show hearing loss, the problem could be conductive hearing loss, perhaps caused by an obstruction or a middle ear infection. If both methods shows hearing loss, the problem is sensorineural, perhaps caused by damage to the cochlea or auditory nerve.

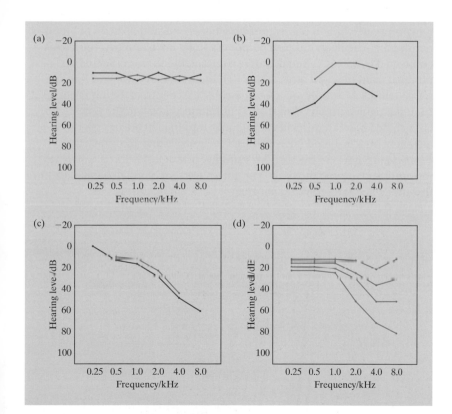

Fig 29
Typical audiograms for: (a) normal ear; (b) conductive hearing loss; (c) sensorineural deafness, and (d) progressive deafness due to exposure to high levels of noise. In (a) to (c), the tests are conducted via headphones (red lines) or using an electronic vibrator held on the bone behind the ear (blue lines). In (d), the curves are for various exposure times (increasing downwards from top to bottom)

Fig 30
Age-related hearing loss. The curves show progressive hearing loss for sounds of different frequencies

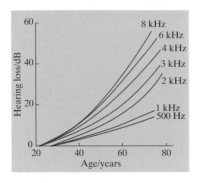

The audiogram of a person who has suffered years of exposure to loud noise is shown in Fig 29(d). Damage may have occurred to the hair cells in the cochlea. Exposure to high-intensity, high-frequency sound is particularly dangerous, though the audiogram always shows maximum loss at 4 kHz, no matter what the frequency of noise that caused the damage.

Age-related hearing loss (Fig 30) tends to affect high-frequency sounds the most. Frequencies below 500 Hz are barely affected. Although human speech is mostly covered by this range, there are higher-frequency harmonics in sounds such as 's' and 't', which older people find difficult to detect.

B.2.3 Biological measurement

Basic structure of the heart

The heart (Fig 31) is a remarkably reliable pump. In an average lifetime, it beats over 2.5 billion times, pumping between 5 and 20 litres of blood per minute. The heart has four chambers – two **atria** and two **ventricles** – connected by valves. When the atria contract, blood from the body is pumped from the right atrium, through the tricuspid valve and into the right ventricle. A short time later, the ventricles contract. Blood is pumped from the right ventricle, through the pulmonary valve, to the lungs. Oxygenated blood returns from the lungs into the left atrium. As the atrium contracts, blood flows through the mitral valve into the left ventricle, from where the blood is pumped through the aortic valve into the aorta and around the body. The valves prevent blood flowing back into the atria when the ventricles contract.

Essential Notes

The pulmonary and aortic valves are collectively referred to as the semi-lunar valves.

Fig 31
The structure of the heart and blood flow

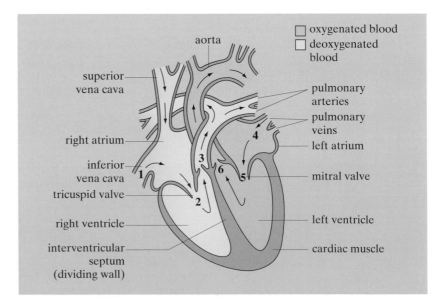

In summary, and referring to the numbers on Fig 31:

1 Blood returning from the body enters the right atrium through the vena cavae.

2 Blood is pumped into the right ventricle.

3 When the right ventricle contracts, it pumps blood through the pulmonary artery to the lungs.

4 Oxygenated blood returns to the left atrium of the heart along the pulmonary veins.

5 Blood is pumped from the left atrium into the left ventricle.

6 The left ventricle contracts, pumping about 70 ml of blood into the aorta and around the body.

Electrical signals and their detection; action potentials

Electrical signals are conducted around the body along nerves. Nerves are made of fibres that can be up to a metre long, but are usually only a few micrometres in diameter. Signals travel along these fibres in the form of a changing potential difference, known as an **action potential** (see below). The action potential is generated by the movement of ions across the cell membrane.

The membrane of the nerve cells allows water to diffuse freely into or out of the cell. However, the membrane is much less permeable to the passage of sodium and potassium ions. It is an imbalance in the numbers of these ions that causes a potential difference across the cell membrane. In all cells, there are certain proteins that act to pump potassium into the cell and sodium out of it (Fig 32). This means that cells have a high potassium concentration and a low sodium concentration, whereas the fluid surrounding the cell has a low potassium concentration and a high sodium concentration. This concentration gradient tends to cause potassium ions to leave the cell, carrying their single positive charges with them. Potassium ions continue to leave the cell until the excess positive charge outside the cell is large enough to stop them. This happens when the inside of the cell is at a potential of −70 mV compared to its surroundings. At this potential difference, an equilibrium exists between the concentration gradient and the potential gradient, and the cell is said to be 'polarised'.

B.2.4 Non-ionising imaging

Ultrasound imaging

Ultrasound is used to produce detailed images in real time at relatively low cost and without risk to the patient. It is suitable for diagnosing a wide range of conditions, from heart valve disorders to tumour detection, as well as for monitoring foetal development.

Generating ultrasound

Ultrasound refers to sound waves of such high frequency that they cannot be detected by the human ear. Strictly speaking, ultrasound starts at about 20 kHz, but medical applications use much higher frequencies, typically between 1 MHz and 20 MHz.

Sound waves are pressure waves that are produced by a vibrating object. The ultrasonic generators used for medicine need to vibrate several million times per second. They use **piezoelectric transducers** (Figs 39 and 40) to convert electrical signals to pressure waves.

Fig 39
The piezoelectric effect.
(a) An a.c. signal is applied across a crystal of piezoelectric material.
(b) One polarity causes the crystal to expand, pushing the air in front of it.
(c) When the polarity reverses, the crystal contracts. (d) This sequence is repeated millions of times each second, causing a pressure wave (ultrasound) to be transmitted

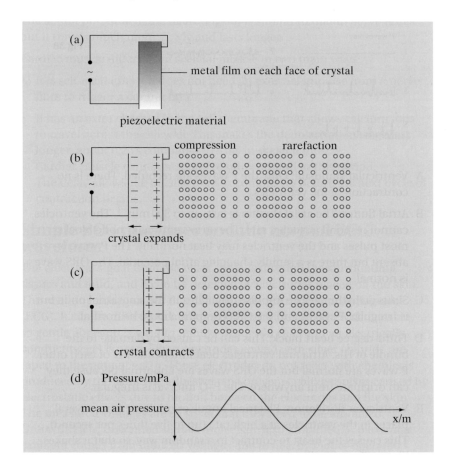

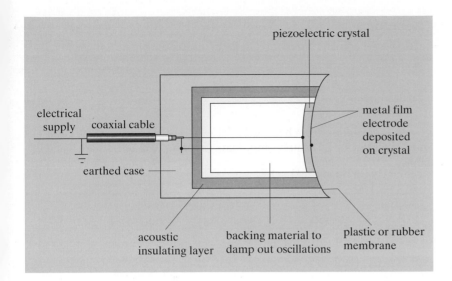

Fig 40
A piezoelectric transducer

Piezoelectric materials deform when a potential difference is applied across them. When an alternating potential difference is applied across the material, it vibrates at the same frequency as the electrical signal. A piezoelectric crystal has a natural frequency that is determined by the material and by its physical dimensions. When the frequency of the a.c. signal matches the natural frequency of the crystal, there is **resonance**: the ultrasonic wave has its largest amplitude for a given input of electrical energy.

The piezoelectric effect also works in reverse. If a piezoelectric crystal is deformed, by an ultrasonic wave for example, a potential difference is generated across the crystal. Such materials can therefore be used to detect, as well as to generate, ultrasound.

Ultrasound travels as longitudinal waves through the body, moving through different tissues at different speeds. The speed of sound, c, in a material depends approximately on the elasticity of the material and its density:

$$c = \sqrt{K/\rho}$$

where K is the bulk elastic modulus and ρ is the average density of the material. The frequency of the sound also has a slight effect on its speed.

As ultrasound travels through the body, some of its energy is reflected as the wave passes from one material to another (see Fig 41, page 137). These 'echoes' from the boundaries between different materials are used to build an image of the internal structure of the body. The amount of energy that is reflected at each interface depends on the **specific acoustic impedance** of the two materials.

Essential Notes

The bulk elastic modulus is similar to the Young modulus, but refers to changes in volume rather than length.

Definition

The specific acoustic impedance, Z, is defined as the product of density, ρ, and the speed of sound, c:

$$Z = \rho c$$

Z has units of $kg\,m^{-2}\,s^{-1}$.

Some typical specific acoustic impedances for body materials are listed in Table 2.

Table 2
Some typical specific acoustic impedances

Material	Speed of sound/$m\,s^{-1}$	Average density/$kg\,m^{-3}$	Specific acoustic impedance $kg\,m^{-2}\,s^{-1}$
bone	3500	1.85	6.48×10^3
fat	1450	0.952	1.38×10^6
muscle	1580	1.08	1.70×10^6

Example

Olive oil has a density of $920\,kg\,m^{-3}$ and a bulk elastic modulus of $1.6\,GPa$. Find the specific acoustic impedance of olive oil.

Answer

From the equations $Z = \rho c$ and $c = \sqrt{K/\rho}$ above, the specific acoustic impedance is

$$Z = \rho c = \rho\sqrt{K/\rho}$$

$$Z = 920 \times \sqrt{\frac{1.6 \times 10^9}{920}} = 1.2 \times 10^6 \; kg\,m^{-2}\,s^{-1}$$

When a plane wavefront strikes the boundary between two materials, at normal incidence (Fig 41), the proportion of the incident intensity, I_I, that is reflected is given by

$$\alpha = \frac{I_R}{I_I} = \left(\frac{Z_2 - Z_1}{Z_2 + Z_1}\right)^2$$

Essential Notes

. The intensity, I, of an ultrasound wave is the energy that is transferred through a unit area per second. The unit of intensity is the watt per square metre ($W\,m^{-2}$).

where I_R is the intensity of the reflected wave, and Z_1 and Z_2 are the specific acoustic impedances of the two media. This formula only applies for normal incidence. The ratio I_R/I_I is known as the **reflection coefficient**, α. The intensity of the transmitted wave, I_T, is the difference between the intensities of the incident and reflected waves:

$$I_T = I_I - I_R$$

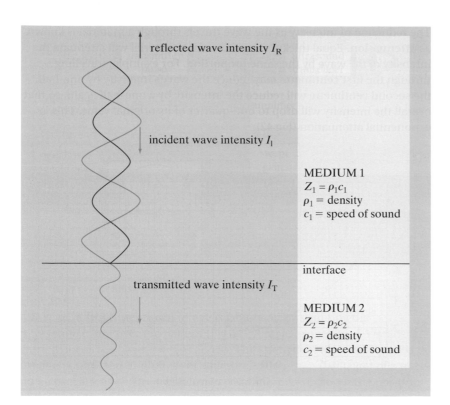

Fig 41
Reflection of ultrasound at a boundary.

Example

The intensity of a diagnostic ultrasound wave is $30\,\mathrm{mW\,cm^{-2}}$.

(a) Find the reflection coefficient for the ultrasound wave as it strikes the boundary between fat and muscle, at normal incidence.

(b) Find the intensity transmitted into the muscle.

Answer

(a) Using the formula in the text and substituting values from Table 2 gives

$$\alpha = \frac{I_R}{I_I} = \left(\frac{Z_{muscle} - Z_{fat}}{Z_{muscle} + Z_{fat}}\right)^2 = \left(\frac{1.70\times10^6 - 1.38\times10^6}{1.70\times10^6 + 1.38\times10^0}\right)^2 = \left(\frac{0.32}{3.08}\right)^2 = 0.011$$

(b) The value found in part (a) gives the reflected intensity as

$$I_R = \alpha I_I = 0.011 \times 30 = 0.33\,\mathrm{W\,cm^{-2}}$$

So the transmitted intensity is

$$I_T = I_I - I_R = 30 - 0.33 = 29.67\,\mathrm{W\,cm^{-2}}$$

As ultrasound travels through a material, its intensity is reduced because the material absorbs some of the wave's energy, transferring it to internal energy. In medical applications, this causes a heating effect in the body tissue. In addition, some of the wave's energy is scattered from its original path, which also reduces the intensity.

The number of complete images, or frames, that can be produced each second depends on how long it takes for the echoes from each pulse to return, and on how many pulses are used to create the full picture. The shortest possible time for one pulse, T_{min}, is given by

$$T_{min} = \frac{distance}{speed} = \frac{2D_{max}}{c} = \frac{2D_{max}}{1500}$$

where D_{max} is the maximum depth that is imaged and c is the speed of sound in tissue ($\approx 1500\,\text{m s}^{-1}$). If there are N lines in the image, N of these A-scans are required and the time for each frame is

$$T_{frame}(s) = \frac{2ND_{max}}{1500}$$

The maximum number of frames per second is the reciprocal of this, i.e.

$$\text{maximum frame rate (Hz)} = \frac{1500}{2ND_{max}}$$

Essential Notes

The higher the frequency, the better the resolution. But higher frequencies are attenuated more. There is a compromise between image resolution and the depth of tissue that can be examined.

There is a compromise between the number of A-scans (lines) in each image and the number of frames per second. More A-scans will improve the resolution of the image, but will limit the frame rate.

Ultrasound scans of the developing foetus are used to determine gestational age and to detect multiple pregnancies or foetal abnormalities. Echocardiography uses ultrasound to provide a real-time image of the heart. For high-resolution images, high-frequency ultrasound must be used, which is 2 MHz for adults and 7 MHz for infants.

Advantages and disadvantages of ultrasound for diagnostic imaging

Ultrasound imaging is non-invasive and causes very little discomfort for the patient. A gel has to be applied to the patient's skin to couple the ultrasound probe to the body, because an air gap would lead to a large reflection and consequent loss of energy. Ultrasound is generally better than X-rays at imaging soft tissue, though structures behind the lung or behind bone cannot be seen, due to the large reflections. At the low energies used in diagnostic ultrasound, there are no known hazards to the patient, or to the operator. Ultrasound equipment is relatively inexpensive, as compared to MR or CT scanning apparatus, and much more portable. However, the images do need a skilful operator to produce them and they require expert interpretation.

Essential Notes

MR stands for magnetic resonance and CT stands for computerised tomography, which uses a series of X-rays to generate an image.

Fibre optics and endoscopy

Modern keyhole surgery uses video cameras to enable the surgeon to see the operation on a screen. Light is carried in to and out of the body by bundles of optical fibres. Optical fibres are very fine, flexible strands of glass that carry light on its twisting route into the body using total internal reflection. Rays of light that strike the inside surface of a glass fibre at less than the **critical angle** are refracted out of the fibre. Any rays that are incident at greater angles are reflected along the fibre just as if the fibre had mirrored internal walls (Fig 45).

Essential Notes

The critical angle c for rays going from medium 1 to medium 2 is given by $\sin c = n_2/n_1$ (see Unit 2).

Fig 45
Total internal reflection in an optical fibre

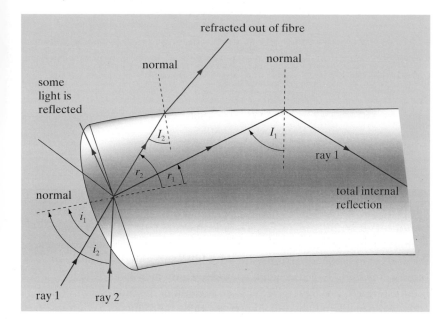

In an **endoscope** (see below), thousands of fine optical fibres are packed together into bundles. If the fibres touch each other, light leaks from one fibre to the next. To prevent this, the fibres are coated with a second layer of glass of slightly lower refractive index. The use of this **cladding** increases the critical angle at the edge of the fibre. The cladding also serves to protect the core from surface scratches, which would also lead to a loss of light.

Example

An optical fibre has a core of refractive index 1.6 and it is coated (cladded) with glass of refractive index 1.5. Find the critical angle at the interface between the glass core and the cladding.

Answer

The critical angle, c, is given by

$$\sin c = \frac{n_{\text{cladding}}}{n_{\text{core}}} = \frac{1.5}{1.6} = 0.9375$$

which gives a value for c of almost 70°.

Endoscopes

An **endoscope** (Fig 46) is a sophisticated medical instrument commonly used to view the gastrointestinal tract. In addition to bundles of optical fibres, it also carries air, water and suction channels, and can carry tools with which to take samples of tissue for analysis.

Fig 46
The endoscope and its tools

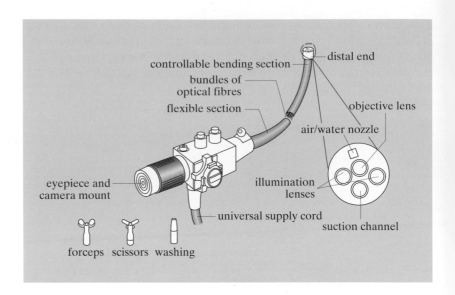

Fig 47
Light is carried down an endoscope in an incoherent bundle, and back up in a coherent bundle

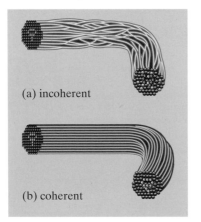

(a) incoherent

(b) coherent

Light is carried to the site of the examination through an **incoherent bundle** of glass fibres. As many as 30 000 individual fibres make up the bundle. An incoherent bundle cannot be used to form an image because the ends of the individual fibres are arranged randomly. In a **coherent bundle**, the fibres have the same spatial position at both ends of the bundle (Fig 47). The light emitted from the end of the bundle is an exact copy of the incident light and an image can be reproduced. Coherent bundles are expensive to manufacture, so incoherent bundles are used for illumination.

Endoscopes are used to examine the upper digestive tract (gastroscopy) or the rectum and colon (colonoscopy). One common application is to look for tumours, without the need for surgical intervention. Arthroscopy is a similar technique used to examine joints, like the shoulder or knee, to diagnose problems such as arthritis. The arthroscope can also be used to treat damaged cartilage or to take small tissue samples (biopsies). Arthroscopy is performed through small incisions. It is much less painful than open surgery, carries less risk of infection and has a faster recovery time.

The laparoscope, a rigid form of endoscope, is used for examining the body through the small incisions made in keyhole surgery. There is often an extra optical fibre used for transmitting laser light, which may be used instead of a scalpel.

MR scanner

MR (magnetic resonance) imaging is used to produce an image of a cross-section through a patient. These images can be used to diagnose a variety of conditions, including cancer. MR scanning requires the patient to be subjected to a strong, uniform magnetic field. In practice, this means that the patient has to lie along the axis of a large solenoid (cylindrical coil).

Although patients may find this claustrophobic, MR scans are considered to be very safe. The patient is not subjected to any ionising radiation as in the case of CT scans, which use X-rays. MR scans produce more detailed images of soft tissue than CT scans do. However, MR scanning takes longer and requires the patient to remain still for longer periods. MR scans are also more expensive than CT scans.

MR scanning relies on the magnetic properties of the proton. A proton forms the nucleus of every hydrogen atom in our bodies. Protons spin, and, because they are charged, they generate their own magnetic field. Normally these fields are randomly orientated. During an MR scan, the patient is placed in a strong magnetic field. The field is produced by large superconducting magnets, which produce a flux density of between 1 and 3 T (tesla) – about 10 000 times greater than the Earth's magnetic field. This strong field acts on the spinning protons, so that their magnetic fields are aligned. A pulse of radio waves is then directed at the patient. This disturbs the orientation of the protons. When the radio pulse stops, the protons return to their original states, emitting a radio-frequency (r.f.) signal as they do so. The strength of the r.f. signal depends on the proton density, which is itself linked to the amount of water in the tissue. It is these signals which are used to reconstruct an image of a section through the patient's body.

Essential Notes

The only significant risk to patients comes from the effect of the strong magnetic field. Patients with a heart pacemaker may not be able to be scanned. There may also be risks for patients with metal implants, though the use of non-ferrous metals, such as titanium, for surgical implants removes this risk.

B.2.5 X-ray imaging

X-rays

X-rays were discovered by Wilhelm Roentgen in 1895. Within a month of the discovery, X-rays were being used by doctors to examine patients with fractures or gunshot wounds. At that time, scientists were unaware of the dangers of ionising radiation, and many of the early patients suffered from high X-ray doses. Modern techniques have made the medical use of X-rays much safer.

Roentgen was experimenting with high-voltage discharge tubes when he discovered X-rays (Fig 48). He used an induction coil to apply a high potential difference, about 35 000 V, between two electrodes in an evacuated glass tube. Electrons released by the cathode were accelerated across the tube towards the anode. The evacuated tube meant that the electrons could reach high speeds, as they were not impeded by collisions with air molecules. Roentgen noticed that a fluorescent screen placed a few metres from the tube produced light when the tube was operating. The fluorescent glow persisted even when Roentgen shielded his tube with thick black paper. He concluded that invisible, penetrating rays were being produced by the tube. He named these X-rays.

We now know that X-rays are very high-frequency electromagnetic waves produced when high-energy electrons lose energy as they collide with atoms.

Fig 48
Roentgen's X-ray tube apparatus

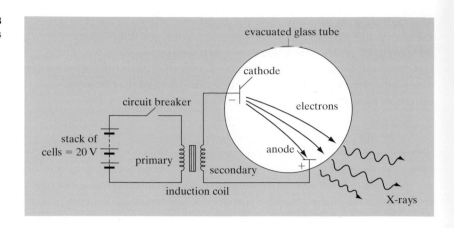

Physical principles of the production of X-rays

In a modern X-ray tube (Fig 49), the cathode is heated by an electric current. The cathode then emits electrons by **thermionic emission**. The electrodes are accelerated towards the anode by a potential difference of around 150 kV. When the electrons reach the target (Fig 50), they collide with the target atoms. As the electrons slow down, they emit a continuous spectrum of X-rays known as **bremsstrahlung** or 'braking' radiation.

Fig 49
A modern X-ray tube has a rotating anode and is mounted in a lead-shielded, oil-cooled case. The anode rotates to reduce local heating problems. The cathode has a heated filament

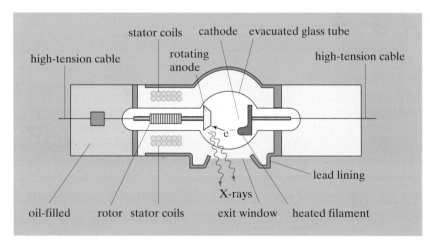

Fig 50
The tungsten target is mounted on the anode. The anode has a bevelled edge to reduce the size of the focal spot. The width of the target is L, but the width of the X-ray beam at the target is $L \sin \theta$. For most applications, θ is about 17°

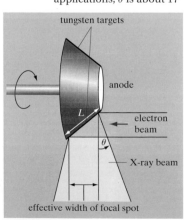

Some of the high-speed electrons will collide with electrons in the target atoms. If they have sufficient energy they will knock atomic electrons out of their orbits and ionise atoms. As a result, **characteristic X-rays** are emitted as other electrons in the target atoms drop from higher energy levels to fill the vacancy (Fig 51). This means that a line spectrum is superimposed on the continuous bremsstrahlung spectrum (Fig 52).

The maximum possible energy of an X-ray photon depends on the potential difference, V, across the tube. As an electron is accelerated, it gains energy, $E = eV$, where e is the charge of the electron. If the electron were to lose all of this energy in one collision with a target atom, the X-ray would be emitted with energy E. The energy of a photon is given by $E = hf$, where h is the Planck constant and f is the frequency of the photon. The wavelength of the emitted photon is therefore

$$\lambda = \frac{c}{f} = \frac{hc}{E} = \frac{hc}{eV}$$

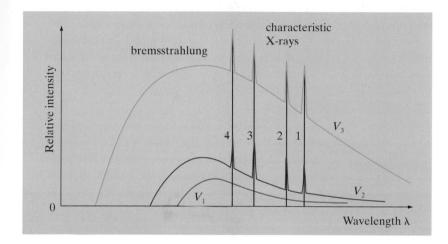

Fig 51

Characteristic X-rays are emitted as electrons fall from higher energy levels to fill any vacancies. The wavelengths emitted depend on the spacing of the energy levels in the target atoms and are therefore characteristic of the target

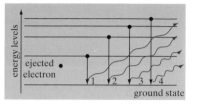

Fig 52

The shape of an X-ray spectrum depends on the potential difference, V, across the tube. Here the three curves are for tube voltages V_1, V_2 and V_3, with $V_3 > V_2 > V_1$. A higher value of V will increase the total intensity emitted (in fact, the total intensity is approximately proportional to V^2). The peak output also shifts to a shorter wavelength, and more characteristic lines may appear

Example

If the potential across an X-ray tube is 150 kV, calculate the minimum X-ray wavelength produced.

Answer

The shortest wavelength of the X-rays, λ_{min}, will be produced if the electron loses all of its energy in one collision. Thus

$$\lambda_{min} = \frac{hc}{eV} = \frac{6.63 \times 10^{-34} \times 3.00 \times 10^8}{1.60 \times 10^{-19} \times 150 \times 10^3} = 8.29 \times 10^{-12} \text{ m}$$

In practice, less than 1% of the energy in the electron beam is transferred to X-ray radiation. The rest is transferred to internal energy in the target, raising its temperature. The target has to be made of a metal with high values of thermal conductivity, specific heat capacity and melting point. Tungsten meets these criteria well and is often used as the target in diagnostic X-ray tubes. Tungsten has a high atomic number, so a tungsten nucleus has a large mass and a high positive charge. This increases the probability that collisions with the high-energy electrons will lead to X-ray emission.

The tungsten target is mounted on the anode, which is able to rotate (see Fig 50). This means that the electron beam strikes different areas of the target and reduces local heating problems. The anode has a bevelled edge so that the target can be wide without increasing the width of the X-ray beam.

Essential Notes

Most of the energy of the X-ray beam will be converted to internal energy in the target. The target needs to be able to conduct this heat away. A large specific heat capacity limits the temperature rise.

The intensity of the X-ray beam is the total energy (emitted at all wavelengths) per second through unit area. (It is the area under the curve in Fig 52.) The X-ray intensity is controlled by the number of electrons striking the target every second, since more electrons leads to more collisions, which means that more X-rays are produced. Therefore, the intensity of the X-ray beam depends on the tube current, which can be controlled by varying the current to the cathode.

Radiographic image detection

Traditionally, X-ray images were detected using photographic film. X-ray photons cause **ionisation** in silver halide grains in the film. When the film is developed, those grains that were exposed to X-rays turn black. The greater the exposure, the more grains that are developed and the darker the film. The variation in darkness (or optical density) between areas of the film below soft tissue, as compared to those below bone, is the **contrast** of the image (Fig 53). Greater contrast will allow doctors to identify abnormalities more easily.

Photographic film is not very sensitive to X-ray radiation. A typical film will absorb less than 0.10% of the X-ray energy that reaches it. Consequently, a relatively long exposure time is needed to get a well-exposed image. But increasing the exposure time increases the dose to the patient, as well as increasing the blurring due to movement. An **intensifying screen** is used to speed things up (Fig 54).

Fig 53

The darkness (or optical density) of the X-ray image gives us information about the attenuation of X-rays in the material above the film. Contrast is the difference in optical density, $OD_1 - OD_2$

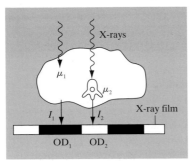

Fig 54

Structure of an intensifying cassette, used to reduce exposure time. X-rays are absorbed by the phosphor which emits blue light

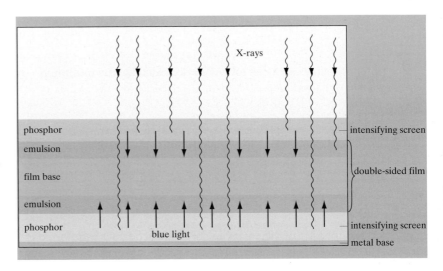

The photographic film is sandwiched between two layers of phosphor. The phosphor, e.g. calcium tungstate, has a high atomic number and therefore absorbs X-rays well. The phosphor emits blue light, which is detected by the photographic film. The overall efficiency of detection is much higher than for film alone. The exposure time, and therefore the dose to the patient, can be reduced.

Doctors often need to see moving images to aid diagnosis, perhaps to see the heart in motion, for example. **Fluoroscopy** uses a fluorescent screen instead of a film. This means that the image can be produced in real time.

An **image intensifier** is used to produce the real-time X-ray image of the patient (Fig 55). X-rays strike a phosphor, often caesium iodide, which then emits light. The light photons strike a photocathode, releasing electrons, which are focused by electrodes inside a vacuum tube. The electrons accelerate across the tube until they strike a fluorescent screen, which provides a visible image, in the same way that a TV screen does. The output can be recorded via a video camera.

The output screen is smaller than the input screen. This reduction, and the acceleration of the electrons across the tube, makes an image up to 1000 times brighter than the original. This means that much lower-intensity X-rays can be used, reducing the dose to the patient. Even with the use of an image intensifier, fluoroscopy gives a relatively high dose, often 15 times that of a conventional X-ray image.

Modern X-ray images are generated digitally using electronic detectors rather than film. This avoids the chemical development needed for films, hence saving time and materials. Digital images can be processed to enhance contrast, without increasing the X-ray dose to the patient. Digital images can be communicated quickly to medical staff, and storage is much easier than for bulky X-ray films.

Image sharpness and contrast

X-ray images are essentially shadow pictures and, like all shadows, the edges are sometimes blurred. The amount of blurring, or **unsharpness**, depends on the size of the X-ray source (Fig 56).

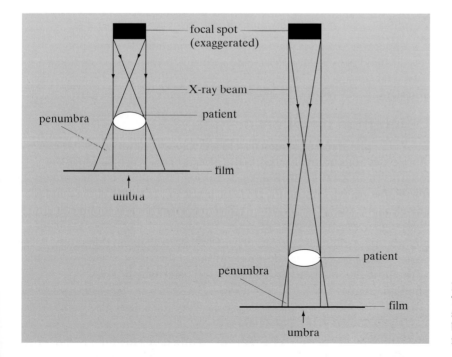

To keep the picture sharp, the size of the X-ray source has to be kept as small as possible. This is done by focusing the electron beam and using a target inclined to the beam. It is also important to keep the film close to

Fig 55
An X-ray image intensifier

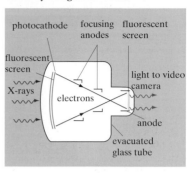

Essential Notes

The most common form of electronic detector uses a silicon detector linked to a scintillator. This gives a flash of light when an X-ray is detected. The light is detected by a **charge coupled device** (**CCD**), which produces a digital image.

Fig 56
The umbra–penumbra effect. The sharpness of the shadow is improved by moving the X-ray source further from the patient

Fig 57
An anti-scatter grid increases contrast by blocking scattered radiation, but some primary radiation is also blocked. A higher exposure is needed, which increases the dose to the patient

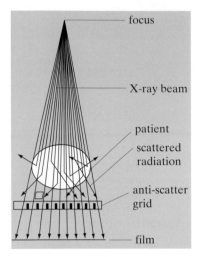

the patient. Unsharpness can be decreased by moving the X-ray source further from the patient, but the inverse square law means that the intensity of the X-ray radiation at the patient is also decreased.

The contrast of an X-ray photograph is reduced by scattered radiation. The amount of scattered radiation can be cut down by placing a grid, constructed from lead strips, between the patient and the film (Fig 57). The grid allows most of the primary radiation to pass through but blocks some of the scattered radiation. The grid itself would be imaged on the film, but moving the grid during the exposure blurs out its shadow.

Reducing the dose

Diagnostic X-rays are responsible for most of our exposure to man-made radiation. It is important to keep the X-ray dose for any particular procedure as small as possible. The dose to the patient depends on the exposure time and the intensity of the beam. The exposure time can be reduced by increasing the sensitivity of the detector, by using an image intensifier or an electronic detector, for example. Using a high tube voltage to produce more penetrating, shorter-wavelength X-rays can produce high-contrast images and allow the tube current (hence intensity) to be lower.

Another way of reducing the dose is to filter out the low-energy component of the X-ray spectrum. Because low-energy photons are unlikely to reach the X-ray film, they cause a dose to the patient without providing any diagnostic information. A filter made from a few millimetres of aluminium is used.

Some organs are more sensitive to radiation than others, so suitable shielding, a lead apron for example, is used to cover these organs.

It is important to reduce the dose to the radiographer. Radiographers should stand behind suitable shielding during the exposure. The inverse square law means that increasing their distance from the X-ray tube will also significantly reduce the dose.

Differential tissue absorption of X-rays

X-rays interact with matter in a number of ways: they can be scattered by atomic electrons or totally absorbed. These processes act to reduce the intensity of an X-ray beam as it passes through the material. The amount of attenuation depends on the atomic number and density of the material, as well as the energy of the photon. Bone is denser than soft tissue. It also has a higher effective atomic number (because bone contains significant amounts of calcium and phosphorus). These factors mean that bone absorbs more X-ray energy than an equivalent thickness of soft tissue. This allows us to differentiate between bone and soft tissue in X-ray photographs.

Exponential attenuation

The reduction in intensity as X-rays are absorbed in a medium is known as **attenuation**. For a narrow, **monoenergetic** X-ray beam, the attenuation is exponential. The intensity I of the beam after passing through a thickness x of a material is given by

$$I = I_0 \, e^{-\mu x}$$

where I_0 is the original intensity of the beam and μ is the **linear attenuation coefficient**, measured in m^{-1}. A given thickness of the material will reduce the X-ray intensity by half. This thickness is referred to as the **half-value thickness**.

We can use the equation $I = I_0\, e^{-\mu x}$ to calculate the half-value thickness (HVT) for bone. We have

$$\frac{I}{I_0} = e^{-\mu x}$$

At the half-value thickness, $x_{1/2}$, we have

$$\frac{I}{I_0} = \frac{1}{2}$$

so we can write

$$\ln\left(\frac{1}{2}\right) = -\mu x_{1/2}$$

$$x_{1/2} = \frac{-\ln\left(\frac{1}{2}\right)}{\mu} = \frac{\ln 2}{\mu}$$

For bone, the linear attenuation coefficient for monoenergetic $150\,\text{keV}$ X-rays is about $0.6\,\text{cm}^{-1}$. So, for bone, HVT is

$$x_{1/2} = \frac{\ln 2}{0.6} = 1.1\,\text{cm}$$

The density of the material, ρ, also affects its ability to absorb energy from X-rays, simply because the X-ray photon encounters more atoms (or more massive atoms) in the same volume. The **mass attenuation coefficient**, μ_m, takes into account the density of the material:

$$\mu_m = \frac{\mu}{\rho}$$

This is now independent of density, so that the value of μ_m is the same for water whether it is in liquid, solid or gaseous form.

Image contrast enhancement

The contrast between images of different soft tissue, or between normal tissue and a tumour, is often very small. One way to visualise internal organs is by using a **contrast medium**. Patients who need an X-ray of their digestive tract often eat a 'barium meal', a thick suspension of barium sulphate.

CT scanner

Conventional X-ray images are shadow pictures, two-dimensional projections of a three-dimensional object, so they carry no information about depth. The limited contrast, especially between soft tissues, is also a problem. X-ray computer tomography (CT) offers a solution to both of these problems. Modern **CT scanners** can produce high-contrast images of a cross-section through the head or body. CT scanning uses a narrow X-ray beam, which is scanned across the patient. A detector records the intensity at each position. The results are digitised and the image is generated by computer. The latest third-generation CT scanners work slightly differently (Fig 58).

Essential Notes

Exponential attenuation of ultrasound is shown in Fig 42.

Fig 58
The set-up of a third-generation CT scanner. The X-ray tube emits a finely collimated fan-shape beam of almost monoenergetic X-rays towards the patient. The X-rays are then detected by an array of detectors. The tube and array is rotated, often through 360°, around the patient

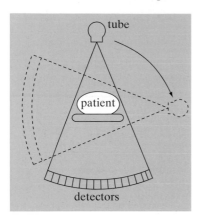

Image contrast depends on different intensities of X-radiation reaching the detector. This in turn depends on the average linear attenuation coefficient in the path of the X-ray beam. We can calculate the intensity reaching the detector, using the equation $I = I_0 e^{-\mu x}$. For a typical diagnostic X-ray beam, muscle has a linear attenuation coefficient of about $0.180\,\mathrm{m}^{-1}$. For each centimetre of muscle:

$$\frac{I}{I_0} = e^{-\mu x} = e^{-0.180 \times 1} = 0.835$$

Similar calculations for bone and blood give values for I/I_0 of 0.619 and 0.837, respectively. So, the contrast between bone and muscle is significant, but the difference between blood and muscle is small, around 0.2%. Contrast media containing iodine may be used intravenously to highlight structures such as blood vessels.

In practice, the X-ray beam passes through a number of different structures and tissues as it travels through the body. The intensity recorded by the detector depends on the thickness and the value of μ of each different material. The image is reconstructed by analysing the intensity measured in many different directions (Fig 59).

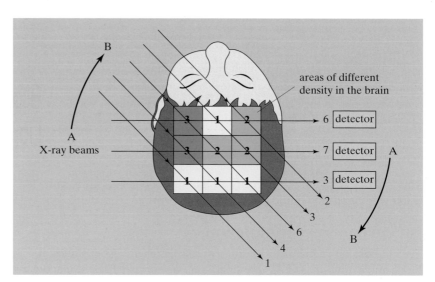

Fig 59
CT scanning

CT scanners are relatively expensive and they give a significantly higher radiation dose to the patient, as compared to a traditional X-ray. These disadvantages are often outweighed by the medical benefits of a detailed image. Modern scanners can reveal detail as small as 1 mm with density differences of less than 1%.

Comparison of imaging techniques

Ultrasound, MR and CT scanning have their own advantages and disadvantages. The main points are summarised in Table 3. The decision as to which technique is most appropriate is a clinical one, and sometimes a diagnosis may demand the use of more than one imaging technique.

Table 3
Advantages and disadvantages of the three scanning techniques

	CT scan	MR scan	Ultrasound
Imaging method	Absorption of X-rays	Magnetic fields and radio waves	Reflected sound waves
Radiation exposure	Moderate to high exposure to ionising radiation	None	None
Risk	Exposure to ionising radiation increases risk of cancer, especially in the foetus	Exposure to strong magnetic fields, but no apparent risk	Low-intensity sound used for diagnosis has no known risk
Time for complete scan	Usually completed within 5 minutes	Scanning can take around 30 minutes	Not used for whole head or body scanning. Real-time imaging of moving parts, e.g. heart
Resolution of different tissues	Not as good at resolving differences between soft tissues. Need to use contrast agents to which patients may react badly. Excellent for imaging bony structures	Much higher detail in soft tissues. Able to adjust contrast by changing the magnetic field and the radio pulses	Less anatomical detail than MR or CT scans. Useful for imaging soft tissue
Cost	CT scanners are expensive, but cheaper than MR scanners	MR scanners are very expensive. Cost per scan is usually more than CT scans	Much cheaper than either CT or MR scanners
Other comments	Lower scan times reduce motion blur. Less problematic for claustrophobic patients	MRI machines can produce images in any plane without moving the patient	Portable. Can be used with Doppler techniques to measure blood flow

What proportion of the incident intensity of ultrasound is transmitted at a boundary between muscle and bone?

_____

_____

_____ 3 marks

(c) Give one reason why ultrasound is preferred to X-rays for foetal scanning.

_____ 1 mark

Total Marks: 8

4 One of the uses of diagnostic X-rays is to diagnose fractures.

(a) Explain why X-ray photographs are ideal for showing bone, but not as effective at showing the differences between soft tissues.

_____

_____

_____ 2 marks

(b) Explain how the image can be made as sharp as possible.

_____

_____

_____

_____

_____

_____ 3 marks

(c) Lead shielding is often used to protect people from X-rays. The linear attenuation coefficient for lead for 150 keV X-rays is $22.8\,cm^{-1}$. Calculate the half-value thickness of lead for 150 keV X-rays.

_____

_____

_____

_____

_____ 3 marks

Total Marks: 8

Answers, explanations, hints and tips

Question	Answer		Marks
1 (a)	Light rays converged / refracted inwards by cornea.	(1)	
	Light rays converged further by lens.	(1)	
	Light rays meet at retina / image drawn inverted.	(1)	
			3
1 (b) (i)	$\text{power} = \dfrac{1}{f}$	(1)	
	where f is the focal length in metres.	(1)	2
1 (b) (ii)	$\dfrac{1}{f} = \dfrac{1}{u} + \dfrac{1}{v}$		
	$\dfrac{1}{v} = \dfrac{1}{f} - \dfrac{1}{u} = 50 - \dfrac{1}{0.4} = 47.5$	(1)	
	$v = \dfrac{1}{47.5} = 0.021\,\text{m} = 2.1\,\text{cm}$	(1)	2
1 (c)	Claire is focusing the image in front of her retina / the image will be blurred.	(1)	
	Claire is short-sighted / myopic.	(1)	
	Claire needs a diverging lens to reduce the power of her eye.	(1)	3
1 (d)	$\dfrac{1}{u} + \dfrac{1}{v} = \dfrac{1}{0.40} + \dfrac{1}{0.022} = 47.9 = 48\,\text{D}$	(1)	
	so she needs $48 - 50 = -2.0\,\text{D}$	(1)	2
			Total 12
2 (a)	Intensity is the power through unit area	(1)	
	at right angles to the direction of the sound wave.	(1)	2
2 (b)	The ear is more sensitive to some frequencies than others.	(1)	
	Loudness is a subjective quantity (different people's ears respond differently).	(1)	2
2 (c) (i)	The ear has an enormous dynamic range, so you need a log scale to show it on a graph. / The ear is a logarithmic device, so 10 times the intensity gives twice the perceived loudness.	(1)	1
2 (c) (ii)	$105 = 10\log_{10}\left(\dfrac{I}{I_0}\right)$		
	$10.5 = \log_{10}\left(\dfrac{I}{I_0}\right)$		
	$10^{10.5} = \dfrac{I}{I_0}$	(1)	
	$I = 10^{10.5} I_0$		
	$I = 3.162 \times 10^{10} \times 1 \times 10^{-12} = 0.032\,\text{W}\,\text{m}^{-2}$	(1)	2
			Total 7

Question	Answer	Marks
3 (a)	Ultrasound is sound of frequency (much) higher than 20 kHz. It is produced by the vibrations of a piezoelectric crystal. The beam is scanned across patient. Echoes result from changes in density. The reflected pulse is detected by a piezoelectric crystal. The time of reflections is used to control the position of the image. The strength of the reflection is used to control image brightness. Mention of brightness scan (B scan). Scanning made more effective by the use of an array of transducers to detect the beam. <div align="right">(any 4)</div>	4
3 (b)	$$\frac{I_R}{I_I} = \left(\frac{Z_2 - Z_1}{Z_2 + Z_1}\right)^2$$ $$\frac{I_R}{I_I} = \left(\frac{1.70 \times 10^6 - 6.48 \times 10^3}{1.70 \times 10^6 + 6.48 \times 10^3}\right)^2 \quad (1)$$ $$= \left(\frac{1.69352 \times 10^6}{1.70648 \times 10^6}\right)^2 = \frac{2.868}{2.912} \quad (1)$$ so 0.985 of the energy is reflected, leaving 0.015 to be transmitted. (1)	3
3 (c)	Not ionising / no chemical changes induced / much safer. (1)	1
		Total 8
4 (a)	It is a shadow picture. The radiation absorbed depends on density. (1) The density of bone is much different from that of tissue. Soft tissue densities / atomic numbers are very similar. (1)	2
4 (b)	The film should be as close as possible to the patient. Use an intensifying screen to reduce the time of exposure (reduce motion blur). Keep the X-ray beam as narrow as possible by using a bevelled edge anode/target. Move the X-ray source as far as possible from the patient. (any 3) Use a grid between the patient and the film to absorb scattered radiation.	3
4 (c)	$I = I_0 e^{-\mu x}$ so $$\frac{I}{I_0} = e^{-\mu x}$$ At the half-value thickness, $x_{1/2}$, $$\frac{I}{I_0} = \frac{1}{2} = e^{-\mu x_{1/2}} \quad (1)$$ so $$\ln\left(\frac{1}{2}\right) = -\mu x_{1/2} \quad (1)$$ $$x_{1/2} = \frac{\ln 2}{22.8} = 0.03 \text{ cm} \quad (1)$$	3
		Total 8

Option Unit 5C Applied Physics

C.3.1 Rotational dynamics

Essential Notes

To convert degrees to radians, you need to remember that there are 2π radians in a circle. So 2π radians = 360°. From this you can work out that

$$1\,\text{radian} = \frac{360°}{2\pi} = 57.3°$$

For a more complete definition, see Unit 4.

For objects and bodies that move in straight lines, we can apply linear dynamics to obtain quantities such as displacement, velocity, acceleration, mass, force and kinetic energy. When objects are rotating, it is more appropriate to apply rotational dynamics to obtain quantities such as **angular displacement**, **angular velocity** and so on. In rotational dynamics, angles and angular displacements are measured in **radians** (abbreviated rad) rather than degrees.

Definition

When the arc length is equal to the radius, the angle θ is equal to one radian.

Fig 1
A particle P moving in a circular path

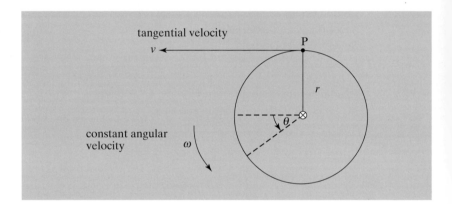

When a particle P moves in a circular path (Fig 1), it has an angular velocity ω given by the angular displacement θ (in radians) of the radius r in time t, i.e.

$$\omega = \frac{\Delta\theta}{\Delta t}\,\text{rad s}^{-1}$$

The tangential velocity v is related to the angular velocity ω by the expression

$$v = r\omega$$

An **angular acceleration** α arises when a particle increases its speed of rotation. It is defined as the rate of change of angular velocity, i.e.

$$\alpha = \frac{\Delta\omega}{\Delta t}\,\text{rad s}^{-2}$$

An increase in angular acceleration must be accompanied by a corresponding increase in the tangential acceleration a, and the relation between these two quantities is

$$a = r\alpha$$

Examiners' Notes

It is important that you can convert between degrees and radians and vice versa.

Concept of moment of inertia

The above dynamics involving movement about a circle considers a body that can be treated as a single particle, where all of its mass is centred at a point. In more realistic situations, such a simplistic assumption cannot be made. The dynamics of a rotating body depend upon the distribution of the mass of the body in relation to a particular **axis of rotation**. The body is effectively composed of an infinite number of point masses (or elemental particles) that move collectively. Each point mass within the body will have a different radius from the axis of rotation and will therefore rotate with a different tangential velocity (since $v = r\omega$).

Consider a two-dimensional **rigid body** (a lamina) that is rotating about some fixed axis, say O, at a constant angular velocity ω, as shown in Fig 2. The rigid body is composed of many point masses such as the one, P, shown with mass m_1, which is at a perpendicular distance r_1 from the axis of rotation.

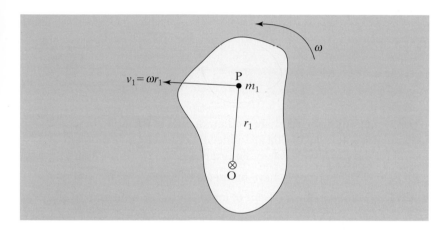

Fig 2
A rigid lamina that is rotating about an axis at O perpendicular to the page

For a particular axis of rotation, there must be a way of measuring the distribution of the body's mass about this axis. The method adopted is to sum the product of each individual point mass with the square of its distance from the point of rotation, i.e.

$$m_1 r_1^2 + m_2 r_2^2 + m_3 r_3^2 + \cdots = \sum_i m_i r_i^2$$

where $i = 1, 2, 3, \ldots$; the mass of the elemental particle i is m_i, and r_i is its perpendicular distance from the axis of rotation. The summation is taken over the whole of the body. This quantity is known as the **moment of inertia**, I.

Essential Notes

The word **inertia** means the resistance of an object to a change in its motion.

Definition

Moment of inertia, $I = \sum mr^2$

The units of I are $\text{kg}\,\text{m}^2$.

The above expression can be applied to any object. In the case of a rigid body with a mass that is continuous, then the summation is replaced by integration. Such mathematical cases will not be considered here, although their final results will be given.

The precise location of the axis of rotation affects the value of the moment of inertia (Fig 3). Distributing the mass further from the centre of rotation (as in a hoop) increases the moment of inertia.

Fig 3
Some examples of rotational inertias. Moments of inertia are given in terms of mass, M (kilogram), and distance from centre of rotation, R (metre)

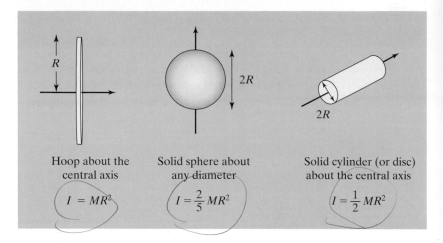

Hoop about the central axis	Solid sphere about any diameter	Solid cylinder (or disc) about the central axis
$I = MR^2$	$I = \frac{2}{5}MR^2$	$I = \frac{1}{2}MR^2$

Examiners' Notes

The expressions for the moments of inertia of different shaped objects will be given in examinations.

Essential Notes

A flywheel is a mechanical device with a large moment of inertia that stores rotational energy.

When two rotating objects share a common axis of rotation, then their individual moments of inertia are added together to give a moment of inertia for the combined system. For example, two flywheels A and B with radii R_A and R_B, respectively, have moments of inertia $\frac{1}{2}M_A R_A^2$ and $\frac{1}{2}M_B R_B^2$. When they combine, the moment of inertia of the combination is

$$I = \frac{1}{2}M_A R_A^2 + \frac{1}{2}M_B R_B^2$$

This holds for any combination provided they rotate about a common axis.

Rotational kinetic energy

Consider a point mass m_1 moving in a circle about some point O a distance r_1 away. The particle is moving with a linear speed v_1. The kinetic energy of the particle is given by

$$E_k = \frac{1}{2}m_1 v_1^2$$

The linear velocity is connected to the angular velocity via the equation $v_1 = r_1\omega$. The kinetic energy can therefore be rewritten as

$$E_k = \frac{1}{2}m_1 (r_1\omega)^2$$

This expression can be rearranged to give

$$E_k = \frac{1}{2}(m_1 r_1^2)\omega^2$$

Comparing this with the first expression for the kinetic energy, it can be seen that if ω is analogous to v then (mr^2) must be analogous to mass m.

An extended body that is rotating about an axis at a point O comprises many such point masses (see Fig 4). The total rotational kinetic energy will then be given by

$$E_k = \frac{1}{2}(m_1 r_1^2)\omega^2 + \frac{1}{2}(m_2 r_2^2)\omega^2 + \cdots$$

$$E_k = \frac{1}{2}\omega^2(m_1 r_1^2 + m_2 r_2^2 + \cdots)$$

with $r_1, r_2, \ldots$ being the perpendicular distances from the axis of rotation. For a given body rotating about a given axis, the term $(m_1 r_1^2 + m_2 r_2^2 + \cdots)$ is the moment of inertia, I, so that the **rotational kinetic energy** can then be given by

$$E_k = \frac{1}{2}I\omega^2$$

When a body is rotating about a centre of rotation, it possesses energy due to its rotational movement. This energy is a result of its motion, not its position, and hence it is rotational kinetic energy, which is distinct from kinetic energy as a result of translational movement.

Fig 4
Calculating the rotational kinetic energy of an extended rotating body

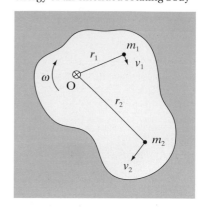

Example 1

A solid steel rotor is being tested. The rotor has a mass of 272 kg and a radius of 38.0 cm. During the test, the rotor reaches an angular speed of 14 000 rev min^{-1} before failing. If the moment of inertia of the rotor is $I = \frac{1}{2}MR^2$, find how much energy was released when the system failed.

Answer

We have

$$I = \frac{1}{2}MR^2 = \frac{1}{2} \times 272 \times 0.38^2 = 19.6\,\text{kg m}^2$$

The angular speed is

$$\omega = \frac{14\,000 \times 2\pi}{60} = 1.47 \times 10^3\,\text{rad s}^{-1}$$

The rotational kinetic energy is

$$E_k = \frac{1}{2}I\omega^2 = \frac{1}{2} \times 19.6 \times (1.47 \times 10^3)^2 = 2.1 \times 10^7\,\text{J}$$

This amount of energy is released when the system fails.

Examiners' Notes

Conversion between rev min^{-1} (or rpm) and rad s^{-1} is a common requirement in questions. One revolution is equivalent to 2π radians, so 100 rev min^{-1} is the same as

$$100\ \text{rpm} = \frac{(100 \times 2\pi)}{60}$$

$$= 10.5\,\text{rad s}^{-1}$$

Example 2

Assuming the Earth to be a sphere with moment of inertia $\frac{2}{5}MR^2$, where M is the mass of the Earth (6.0×10^{24} kg) and R is the radius of the Earth (6.4×10^6 m), determine (a) the value of the moment of inertia of the Earth and (b) the Earth's rotational kinetic energy.

Answer

(a) The moment of inertia is

$$I = \frac{2}{5}MR^2 = \frac{2}{5} \times 6.0 \times 10^{24} \times (6.4 \times 10^6)^2$$
$$I = 9.8 \times 10^{37} \text{ kg m}^2$$

(b) The rotational kinetic energy is

$$E_k = \frac{1}{2}I\omega^2$$

where ω is the angular velocity. For the Earth $T = 24\,\text{h} = 86\,400$ s, and ω is given by

$$\omega = \frac{2\pi}{T} = \frac{2\pi}{86\,400} = 7.3 \times 10^{-5} \text{ rad s}^{-1}$$

Therefore,

$$E_k = \frac{1}{2} \times 9.8 \times 10^{37} \times (7.3 \times 10^{-5})^2$$
$$E_k = 2.6 \times 10^{29} \text{ J}$$

Angular displacement, velocity and acceleration

The equations of motion of linear dynamics have an exact analogy when dealing with rotational dynamics (Table 1). The equations for rotational dynamics are subject to a constant (uniform) angular acceleration α. Angular acceleration is defined as the rate of change of angular velocity (see page 158).

Table 1
The similarity between the linear equations of motion for constant linear acceleration and the rotational equations of motion for constant angular acceleration

Linear equation of motion	Rotational equation of motion
$v = u + at$	$\omega_2 = \omega_1 + \alpha t$
$s = ut + \frac{1}{2}at^2$	$\theta = \omega_1 t + \frac{1}{2}\alpha t^2$
$v^2 = u^2 + 2as$	$\omega_2^2 = \omega_1^2 + 2\alpha\theta$
$s = \frac{1}{2}(u + v)t$	$\theta = \frac{1}{2}(\omega_1 + \omega_2)t$

Examples of the use of the equations of rotational motion are given on pages 164–165.

Torque and angular acceleration

In linear dynamics, Newton's second law states that a resultant force F produces an acceleration a on a mass m, according to the equation

$$F = ma$$

In rotational dynamics, this force is replaced by the **torque** T. The torque is the moment (or turning effect) of a force about a point.

> **Definition**
>
> A torque is the product of a force and the perpendicular distance of the line of action of the force from a particular point of rotation.
>
> The unit of torque is newton metre (N m). The convention adopted is that positive torques produce a clockwise rotation.

So in Fig 5 we have a positive torque T given by

$$T = F \times s = Fs$$

In a situation where the axis of rotation is at the centre of an object or system, there will be two equal and opposite, parallel forces (Fig 6), whose lines of action do not coincide. In such a situation, the forces are known as a **couple** and such couples produce rotational movement.

In this case the torque T is given by

$$T = \left(F \times \frac{s}{2} \right) + \left(F \times \frac{s}{2} \right) = Fs$$

This is identical to the previous expression. The torque produced by a couple is equal to the product of one of the forces and the perpendicular distance between the forces.

A torque therefore produces a rotation and hence an angular acceleration. The action of a torque is to produce an angular acceleration α in the same way as a force produces a linear acceleration according to Newton's second law. The corresponding equation, and hence the rotational counterpart of Newton's second law, can be derived as follows (see Fig 7). For the particle m_1 we have

$$F_1 = m_1 a_1 = m_1 r_1 \alpha \quad \text{(since } a = r\alpha, \text{ see page 158)}$$

The torque on this particle is

$$F_1 r_1 = m_1 r_1^2 \alpha$$

The total torque T on the whole body is the sum of all the individual torques:

$$T = F_1 r_1 + F_2 r_2 + \cdots = m_1 r_1^2 \alpha + m_2 r_2^2 \alpha + \cdots$$
$$= (m_1 r_1^2 + m_2 r_2^2 + \cdots)\alpha$$

Fig 5
The torque about a point ($\otimes$) is the product of the force and the perpendicular distance of the line of action of the force from the point

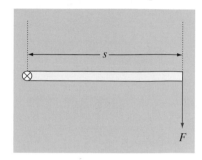

Fig 6
A couple consists of two equal, opposite and parallel forces

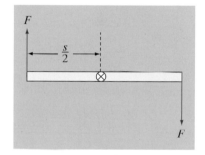

Fig 7
The total torque on an extended body is the sum $F_1 r_1 + F_2 r_2 + \ldots$

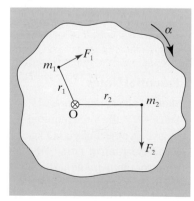

Therefore the total torque T on the whole body is

$$T - \left(\sum mr^2\right)\alpha = I\alpha$$

where I is the moment of inertia.

In this case, the applied torque T has units of N m, the moment of inertia I is in $kg\,m^2$ and the angular acceleration α is in $rad\,s^{-2}$.

Example 1

A flywheel has a moment of inertia $2.4 \times 10^{-2}\,kg\,m^{-2}$ and is rotating with an angular velocity of $20\,rad\,s^{-1}$. Calculate the torque that is required to bring the flywheel to rest in 5 revolutions.

Answer

Given that $I = 2.4 \times 10^{-2}\,kg\,m^{-2}$, $\omega_1 = 20\,rad\,s^{-1}$ and $\theta = 5$ revolutions $= 5 \times 2\pi = 10\pi$ radians, and knowing that we can use the following equation of rotational motion to calculate the angular acceleration:

$$\omega_2^2 = \omega_1^2 + 2\alpha\theta$$

Hence,

$$0 = 400 + 2\alpha(10\pi)$$

giving

$$\alpha = -\frac{20}{\pi}\,rad\,s^{-2} = -6.37\,rad\,s^{-2}$$

i.e. a deceleration.
The torque is then

$$T = I\alpha = -2.4 \times 10^{-2} \times 6.37 = -0.153\,Nm$$

So a torque of $0.153\,N\,m$ opposing the motion is needed.

Example 2

A park roundabout has a diameter of $4\,m$. A small child provides a tangential force of $50\,N$ at the edge of the roundabout, causing an angular acceleration of $0.4\,rad\,s^{-1}$. Determine the moment of inertia of the roundabout.

Answer

Here the radius is half the diameter, so $s = 2\,m$, $F = 50\,N$ and $\alpha = 0.4\,rad\,s^{-2}$.

Torque is force $\times$ distance, so

$$T = Fs = 50 \times 2 = 100\,Nm$$

Using

$$T = I\alpha$$

and rearranging gives the moment of inertia as

$$I = \frac{T}{\alpha} = \frac{100}{0.4} = 250\,kg\,m^2$$

Essential Notes

Here T/α has units $N\,m/rad\,s^{-2}$. Since $1\,N$ is the force required to give an object of mass $1\,kg$ an acceleration of $1\,m\,s^{-2}$, the units for moment of inertia are $(kg\,m\,s^{-2} \times m)/rad\,s^{-2}$, which is $kg\,m^2$.

Example 3

A flywheel is mounted on a horizontal axle of diameter 0.1 m. A constant force of 100 N is applied tangentially to the axle. If the moment of inertia of the whole system (flywheel and axle) is $8\,\text{kg m}^2$, determine
(a) the angular acceleration of the flywheel, and
(b) the number of revolutions the flywheel makes in 30 s (assume the flywheel starts from rest).

Answer

It is a good idea to draw a simple diagram containing the information given (see Fig 8).

(a) We have

$$\text{torque} = \text{force} \times \text{distance}$$
$$T = 100 \times 0.05 = 5\,\text{Nm}$$

We also know that

$$T = I\alpha$$

Hence the angular acceleration is

$$\alpha = \frac{T}{I} = \frac{5}{8} = 0.625\,\text{rad s}^{-2}$$

(b) Initial angular velocity $\omega_1 = 0$, and $t = 30\,\text{s}$. Then the total angle turned, $\theta\,\text{rad}$, is given by the following equation of rotational motion:

$$\theta = \omega_1 t + \tfrac{1}{2}\alpha t^2$$
$$\theta = 0 + \tfrac{1}{2} \times 0.625 \times 30^2 = 281\,\text{rad}$$

Since 1 revolution $= 2\pi$ rad, the total number of turns the flywheel makes is

$$\frac{281}{2\pi} = 44.8\,\text{revolutions} \approx 45\,\text{revolutions}$$

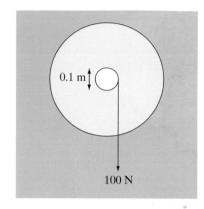

Fig 8
A flywheel mounted on a horizontal axle

Angular momentum

The **angular momentum** L of a particle about an axis is the product of its linear momentum and the perpendicular distance of the particle from the axis, i.e.

$$L = m_1 v_1 \times r_1$$

For a particle P rotating about an axis at O with an angular velocity ω (where $v_1 = r_1\omega$), this leads to the expression

$$L = m_1 \omega r_1^2$$

For a body composed of many such point masses $m_1, m_2, m_3, \ldots$ with distances $r_1, r_2, r_3, \ldots$ from the axis of rotation, the total angular momentum is given by

$$L = m_1 \omega r_1^2 + m_2 \omega r_2^2 + \cdots$$
$$L = \omega(m_1 r_1^2 + m_2 r_2^2 + \cdots)$$

leading to the expression

$$L = I\omega$$

The above equation only applies to a rigid body with a fixed axis of rotation.

Angular momentum is a vector quantity and has an associated direction, which is along the axis of rotation, and hence perpendicular to the plane of rotation (Fig 9). By convention, the direction of the angular momentum vector is towards an observer if the direction of rotation is anticlockwise.

Fig 9
The angular momentum vector is perpendicular to the plane of rotation

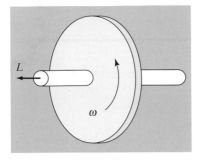

Example

The London Eye has a diameter of 135 m and makes a complete rotation at a constant angular speed ω in 30 minutes. If the total moment of inertia of the Eye is $8.2 \times 10^9 \, \text{kg m}^2$, calculate the magnitude of the angular momentum.

Answer

To find the angular speed, we use the fact that the wheel goes through an angular displacement of $\theta = 2\pi$ rad in a time period $T = 30$ min. Hence,

$$\omega = \frac{2\pi}{30 \times 60} = 0.0035 \text{ rad s}^{-1}$$

Using the expression for angular momentum gives

$$L = I\omega = 8.2 \times 10^9 \times 0.0035 = 2.9 \times 10^7 \text{ kg m}^2 \text{ s}^{-1}$$

Essential Notes

The conservation of linear momentum was considered in Unit 4.

When dealing with linear momentum, we can invoke the conservation of linear momentum of a body provided that no external forces are applied. In a similar way, the total angular momentum at some initial time (i) is equal to the total angular momentum at some time later (f) provided no external torque acts upon the system, no matter what takes place within the system. So

$$L_i = L_f$$

$$I_i \omega_i = I_f \omega_f$$

Essential Notes

Conservation of angular momentum may result in bodies rotating in opposite directions as a result of a collision.

This is called the **conservation of angular momentum**. The law has no exceptions and holds true for a range of physics situations from subatomic particles to those that travel close to the speed of light (areas where Newton's law is not applicable).

Conservation of angular momentum applies to the elliptical motion of planets orbiting the Sun. In these cases, the direction of the gravitational force is perpendicular to the orbital motion, and hence the force exerts no torque on a planet. If the angular momentum of the planet is conserved, then the angular velocity of the planet must increase if its moment of inertia decreases. This has been verified, since planets move faster when closer to the Sun and slower when further away.

Imagine a spinning gyroscope that is fixed in a horizontal position on a vertical pole attached to a turntable that is free to rotate (Fig 10a). The gyroscope will continue spinning in its fixed position and the turntable will remain still (Fig 10a). However, if the spinning gyroscope is now positioned vertically, then the turntable will rotate in the opposite direction to the spinning gyroscope (Fig 10b), so that angular momentum is conserved. For a spinning gyroscope that is not fixed but is on a vertical pivot (Fig 10c), the action of gravity forces the gyroscope to rotate slowly about the pivot point, i.e. it precesses in an anticlockwise direction.

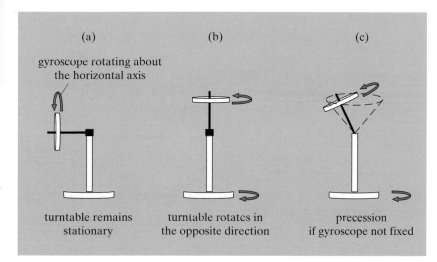

Fig 10
Effects of a spinning gyroscope on a freely rotating turntable

Example

An ice skater is spinning about a vertical axis. Explain what happens when he brings his arms much closer to his body.

Answer

The act of bringing his arms closer to his body means that his mass is much closer to his axis of rotation. This results in a lowering of the moment of inertia I. As angular momentum ($I\omega$) is conserved, this results in a much larger value of ω, the angular velocity, so the skater spins faster. There is a corresponding increase in the rotational kinetic energy $\frac{1}{2}I\omega^2$. This energy has come from the skater, who has had to do work to pull his arms in.

An **angular impulse** is a change in total angular momentum, so

$$\Delta L = L_f - L_i = I\omega_f - I\omega_i$$

It has the same units as angular momentum.

The torque T applied can be related to the angular impulse ΔL. From page 164, the torque is equal to the rate of change of angular momentum:

$$T = I\alpha = \frac{\Delta\omega}{\Delta t} = \frac{\Delta L}{\Delta t}$$

This also applies in the more general case when the moment of inertia I is not constant. The expression

$$\Delta L = T\,\Delta t$$

is analogous to impulse (change in linear momentum) as the product of the average force F and the duration of the impact Δt.

Power

Consider a rigid body that turns through an angle θ about an axis when a force F is applied. The perpendicular distance from the axis to the line of action of the force is s. The force produces a constant torque T given by

$$T = Fs$$

As the rigid body rotates about the axis, the force moves along an arc of length $s\theta$. The **work done** is the product of the force and the distance moved, i.e.

$$W = Fs\theta$$

which is

$$W = T\theta$$

Here, the work done, W, is in joules when the constant torque, T, is in N m and the angle moved, θ, is given in radians.

The work done can be used to overcome any resistive forces that may be present in the rotating system, such as friction. Alternatively, the work done may be used to increase the rotational kinetic energy of the rigid body.

Power is the rate of doing work, i.e.

$$P = \frac{\Delta W}{\Delta t} = \frac{\Delta(T\theta)}{\Delta t} = T\frac{\Delta\theta}{\Delta t}$$

where $\Delta\theta/\Delta t$ is the change in angular displacement with respect to time, i.e. angular velocity. Hence, power is given by

$$P = T\omega$$

Summary comparison

The correspondences between linear and rotational motion are summarised in Table 2.

Linear motion	Linear quantity	Rotational motion	Rotational quantity	Rotational units
mass	m	moment of inertia	I	kg m^{-2}
velocity	$v = \dfrac{\Delta s}{\Delta t}$	angular velocity	$\omega = \dfrac{\Delta \theta}{\Delta t}$	rad s^{-1}
acceleration	$a = \dfrac{\Delta v}{\Delta t}$	angular acceleration	$\alpha = \dfrac{\Delta \omega}{\Delta t}$	rad s^{-2}
displacement	s	angular displacement	θ	rad
translational kinetic energy	$E_T = \dfrac{1}{2}mv^2$	rotational kinetic energy	$E_k = \dfrac{1}{2}I\omega^2$	J
momentum	$p = mv$	angular momentum	$L = I\omega$	N m s
force	$F = ma$ $F = \dfrac{\Delta(mv)}{\Delta t}$	turning moment (torque)	$T = Fs$ $T = I\alpha$ $T = \dfrac{\Delta(I\omega)}{\Delta t}$	N m
work done	$W = Fs$	work done	$W = T\theta$	J
power	$P = Fv$	power	$P = T\omega$	W
impulse	$I = Ft$	angular impulse	$\Delta L = I\omega_f - I\omega_i$ $\Delta L = T\Delta t$	N m s

Table 2
The exact correspondence between linear motion and angular motion

C.3.2 Thermodynamics and engines

Thermodynamics is the study of the relationship between heat and other forms of energy. As such, thermodynamic **equations of state** involve the fundamental properties and intimate connections between the pressure, volume and temperature of gases, vapours and liquids. The fundamental relationships between pressure p, volume V and temperature T are known as the three **gas laws**.

Essential Notes

The gas laws are treated in more detail in Unit 5, section 3.5.3, page 36.

For a fixed mass of gas at a constant temperature, the product of the pressure and volume is constant. This is known as **Boyle's law** and is represented by the equation

$$pV = \text{constant} \quad \text{or} \quad p \propto \frac{1}{V} \text{ at constant } T$$

Essential Notes

The gas laws only apply to a fixed mass of gas.

For a fixed mass of gas at constant pressure, the volume is directly proportional to the temperature measured in kelvin. This is called **Charles' law** and is represented by the equation

$$\frac{V}{T} = \text{constant} \quad \text{or} \quad V \propto T \text{ at constant } p$$

> **Definition**
>
> A temperature change of 1 **kelvin** (1 K) is identical to a temperature change of 1 °C. The kelvin and Celsius scales have equal increments, but they start at different places: 0 K = −273.15 °C and 0 °C = 273.15 K.

Essential Notes

The three gas laws are not independent, since any one of them can be derived from the other two.

For a fixed mass of gas at constant volume, the pressure is directly proportional to the temperature measured in kelvin. This is known as the **pressure–temperature law**, or **pressure law**, and is given by

$$\frac{p}{T} = \text{constant} \quad \text{or} \quad p \propto T \text{ at constant } V$$

No real gas obeys the gas laws exactly. The gas laws apply to the **ideal gas**, and provide an accurate description of the way gases behave at low pressures and at temperatures that are well above their liquefaction temperature. The above three gas laws can be embodied in a single equation, called the **ideal gas equation**, or **equation of state for an ideal gas**, which for *one mole* of gas is written as

$$pV = RT$$

where p is the gas pressure in N m^{-2} or Pa (pascal), V is the gas volume in m^3, T is the temperature in kelvin, and R is called the universal **molar gas constant**, $R = 8.31 \, \text{J K}^{-1} \text{mol}^{-1}$ (here mol is the short form of **mole**). For n moles of a gas, the ideal gas equation becomes

$$pV = nRT$$

A gas that obeys this equation exactly has no forces acting between the molecules, and hence the **internal energy**, i.e. the energy of the molecules, is entirely kinetic and depends only on its temperature.

First law of thermodynamics

The **first law of thermodynamics** relates heat and work, and is essentially the principle of the conservation of energy. It is given by the equation

$$Q = \Delta U + W$$

which applies, for example, to a gas that is being heated, whose temperature increases and which expands, doing work. Here Q is the heat entering the system, ΔU is the increase in internal energy and W is the work done *by* the system. It follows that the internal energy can be increased either by doing work *on* the system (by compressing it) or by putting heat energy into the system. When the internal energy of a system changes, the change depends only on the initial and final states of the system, and not on how the change was brought about.

When a system is *isolated*, it is cut off entirely from any external influence. No work can be done, hence $W = 0$, and no heat can enter or leave the system, hence $Q = 0$. It follows then that $\Delta U = 0$ and the internal energy of an isolated system is constant.

One special type of change that can occur to a system is called an **adiabatic process.** This is one that occurs very rapidly or in a system that is so well insulated that no transfer of energy as heat can occur between the system and its external environment. In this situation, $Q = 0$ and the equation reduces to $\Delta U = -W$. This means that, under adiabatic conditions, any increase in the internal energy of the system is equal to the work done on it.

Non-flow processes

There are four different thermodynamic processes in which there is a certain restriction imposed on the system. The consequences of applying the first law of thermodynamics to these processes are summarised below.

Isothermal processes

> **Definition**
>
> An **isothermal process** is a process that takes place under constant-temperature conditions.

For a gas that approximates to ideal, it follows from the ideal gas equation that

$$pV = \text{constant}$$

The internal energy of an ideal gas depends only on its temperature. For an ideal gas involved in an isothermal process, $\Delta U = 0$, and the first law of thermodynamics reduces to

$$Q - W$$

If a gas expands and does external work W, an amount of heat Q has to be supplied to the gas in order to maintain its temperature, and vice versa. To produce an isothermal change requires a gas to be kept in a thin-walled

Essential Notes

The internal energy of a system is the sum of the kinetic and potential energies of the atoms and molecules comprising the system. The value of the absolute internal energy of a system cannot be measured, only its change from one state into another.

Essential Notes

Internal energy of a system is covered in Unit 5, section 3.5.3, page 44.

Essential Notes

It is important to understand that $+W$ represents work done *by* the system and $-W$ represents work done *on* the system.

Essential Notes

In practice it is not possible to produce a perfectly reversible change.

vessel that is a composed of an excellent conducting material surrounded by a constant-temperature bath. Any expansion or contraction within the system must take place slowly. Under such conditions the process can be reversed and returned to its initial state, and the process is termed **reversible**.

For a reversible isothermal change,

$$p_1 V_1 = p_2 V_2$$

where p_1 and V_1 are the initial pressure and volume of the gas, and p_2 and V_2 are the pressure and volume after the isothermal change.

Adiabatic processes

> **Definition**
>
> An **adiabatic process** is a process that takes place in such a way that no heat can enter or leave the system.

Table 3
The value of γ for gases of different molecular structure

γ	Gas
1.67	monatomic
1.40	diatomic
1.33	polyatomic

When an ideal gas undergoes a reversible adiabatic expansion or contraction, the gas equation used is

$$pV^{\gamma} = \text{constant}$$

where the value of γ depends on the molecular structure of the gas and is given by the ratio of the principal heat capacities, C_P/C_V (see below), of the gas involved.

For an adiabatic process, $Q = 0$, and the first law of thermodynamics reduces to

$$W = -\Delta U$$

If the gas expands and does external work, its temperature falls, and vice versa. A truly adiabatic process is not possible, but processes that happen suddenly and rapidly can produce near-adiabatic conditions, e.g. the rapid expansion and contraction of air when sound waves pass through.

The gas equation for an adiabatic change can also be given as

$$p_1 V_1^{\gamma} = p_2 V_2^{\gamma}$$

Essential Notes

Heat capacity is treated in more detail in Unit 5, section 3.5.3, page 31.

where p_1 and V_1 are the initial pressure and volume of the gas, and p_2 and V_2 are the pressure and volume after the adiabatic change.

Since the ideal gas equation applies to any change of state, it can expressed in the form

$$\frac{p_1 V_1}{T_1} = \frac{p_2 V_2}{T_2}$$

Combining these two equations eliminates the pressure, giving the equation

$$T_1 V_1^{(\gamma-1)} = T_2 V_2^{(\gamma-1)}$$

which allows the final temperature to be calculated.

Constant-volume processes

If the volume of a system such as a gas is held constant, i.e. we have a **constant-volume process**, then the system can do no work, i.e. $W = 0$. The first law of thermodynamics then reduces to

$$\Delta U = Q$$

If heat is absorbed by a system (i.e. Q is positive), the internal energy of the system must increase, and vice versa.

Constant-pressure processes

If, however, the pressure p of the system such as a gas is held constant and the gas expands, then the external work done, W, by the expansion is

$$W = pA\,\Delta s$$

where A is the cross-sectional area and Δs is the small distance moved (e.g. by the piston). In other words,

$$W = p\,\Delta V = p\left(V_f - V_i\right)$$

where ΔV is the small increase in volume of the gas. The first law of thermodynamics then gives

$$Q = \Delta U + p\,\Delta V$$

A summary of these special cases is given in Table 4.

Essential Notes

Under these conditions $Q = C_V \Delta T$, where C_V, the molar heat capacity of a gas at constant volume, is the heat required to produce unit temperature rise in one mole of the gas when the volume remains constant, and ΔT is the temperature rise.

Essential Notes

Under these conditions $Q = C_p \Delta T$, where C_p, the molar heat capacity of a gas at constant pressure, is the heat required to produce unit temperature rise in one mole of the gas when the pressure remains constant, and ΔT is the temperature rise.

First law of thermodynamics: Q = ΔU + W		
Process	**Restriction**	**Consequence**
isothermal	$\Delta U = 0$	$Q = W$
adiabatic	$Q - 0$	$W = -\Delta U$
constant volume	$W - 0$	$\Delta U = Q$
constant pressure	$W = p\,\Delta V$	$Q = \Delta U + p\,\Delta V$

Table 4
A summary of the special cases of the first law of thermodynamics

The *p–V* diagram

The plot in Fig 11 of the pressure versus the volume of a gas is called a **p–V diagram.** It can be shown that the work done, W, by the gas in changing its volume from an initial state V_i to a final state V_f is given by

$$W = \int_{V_i}^{V_f} p\,dV = \text{area of shaded region below curve}$$

Essential Notes

A *p–V* diagram is also known as an **indicator diagram**.

Fig 11

A p–V diagram showing the expansion of a gas. The work done *by* the gas is the area under the curve

If the arrow is reversed the diagram will represent the compression of a gas. The work done by the gas $W < 0$. The area under the curve is the work done *on* the gas

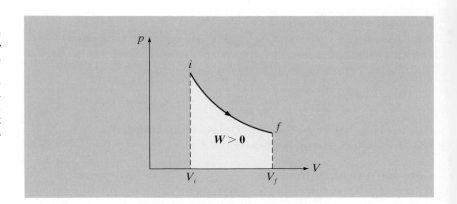

Some more p–V diagrams are shown in the following figures, for different situations. In the **cyclic process** in Fig 12, the work done per cycle is given by

work done per cycle = area of loop

Fig 12

A p–V diagram showing a thermodynamic cycle in which the system is taken back to its original state. The net work is positive, since the area under the expansion curve is greater than the area under the compression curve

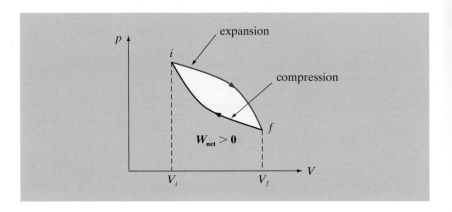

Fig 13 compares an isothermal expansion and an adiabatic expansion of an ideal gas from the same initial state. The adiabatic expansion results in a fall in temperature.

Fig 13

A comparison of an isothermal and an adiabatic expansion. The work under the curve is greater for the isothermal expansion than for the adiabatic expansion, i.e. more work is done. Note also that the adiabatic curve is steeper at all points ($\gamma > 1$)

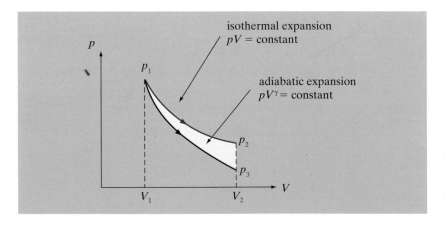

The p–V diagram in Fig 14 shows an adiabatic expansion followed by an isothermal compression.

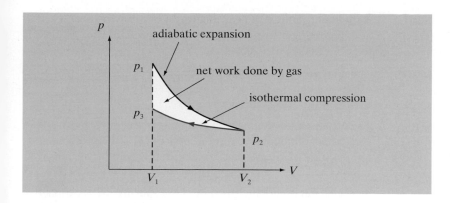

Fig 14
A *p–V* diagram of a gas that has been expanded adiabatically from volume V_1 to V_2 and then compressed isothermally back to its original volume V_1. The net work done by the gas is the shaded region between the curves

Engine cycles

A device or system that extracts energy from its environment in the form of heat and converts it into useful work is called a **heat engine**. At the heart of every heat engine is a working substance. Typical examples include the internal combustion engine, which uses a petrol–air mixture, and the steam engine, which uses water. For such engines to be useful, they must work continuously and work in a cycle, i.e. the working substance must pass through a series of thermodynamic processes.

Two types of engine cycle will be looked at in some detail: the four-stroke petrol engine cycle and the Diesel engine cycle.

Idealised four-stroke petrol engine cycle
Fig 15 shows the *p–V* diagram for the cycle of an idealised petrol engine.

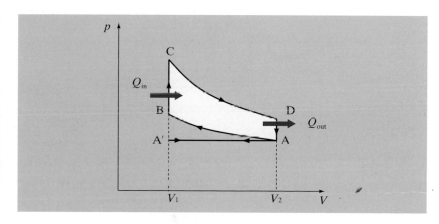

Fig 15
The *p–V* diagram for an idealised four-stroke petrol engine cycle. The diagram for an actual engine has rounded corners

Essential Notes

The idealised four-stroke petrol engine cycle is often referred to as the Otto cycle.

The following is the sequence of processes that take place.

A′ to A   The inlet valve opens and the exhaust valve closes. The piston moves down. During this time, a mixture of typically 7% petrol vapour and 93% air at about 50 °C is drawn into each cylinder. This is the *induction stroke*.

At A   The inlet valve closes.

A to B   The piston moves up, compressing the gas adiabatically. The temperature rises to 300 °C. This is the *compression stroke*.

B to C   A spark plug ignites the gas mixture at B supplying heat (Q_{in}) and increasing the pressure at constant volume. The temperature rises to 2000 °C.

C to D   The increased pressure pushes the piston down as the gas expands adiabatically, decreasing both the pressure and temperature. This is the *power stroke*.

D to A   The exhaust valve opens at D and most of the burnt gas mixture is released, removing an amount of heat Q_{out}. The pressure and temperature of the gas that remains in the cylinder decrease rapidly.

A to A'   As the piston moves up, the remaining gas mixture is expelled. This is the *exhaust stroke*.

At A'   The exhaust valve closes and the inlet valve opens. The cycle repeats.

Essential Notes

Efficiency is a measure of the performance of an engine. The thermal efficiency (see page 180) is the ratio of the work done by the engine to the heat supplied by the fuel.

The work done by the engine in one cycle is the area ABCD shown shaded in Fig 15.

The efficiency of an actual engine is considerably less than that predicted on theoretical grounds, for several reasons. The petrol–air mixture does not behave as an ideal gas, and the efficiency is further reduced by frictional effects, turbulence, loss of heat to the cylinder walls, and the fact that the inlet and exhaust valves take a finite time to open and close. Typical actual efficiencies are ~28% compared to a theoretical efficiency of ~58%.

Idealised Diesel engine cycle

Fig 16 shows the p–V diagram for the cycle of an idealised Diesel engine.

Fig 16
The p–V diagram for an idealised Diesel cycle. The diagram for an actual engine has rounded corners

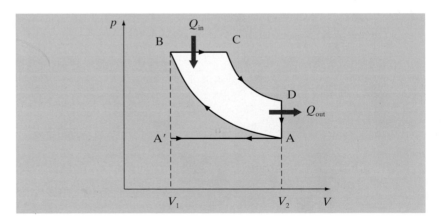

The following is the sequence of processes that take place.

At A'   The inlet valve opens and the exhaust valve closes.

A' to A   Air is drawn into each of the cylinders at atmospheric pressure as the piston moves down. This is the *induction stroke*.

At A   The inlet valve closes.

A to B   The air in the cylinder is compressed adiabatically as the piston moves up. The temperature of the air rises to 700 °C and becomes hot enough to ignite the fuel. This is called the *compression stroke*.

B to C   Diesel fuel is sprayed into the cylinder and is ignited immediately by the hot air, supplying heat (Q_{in}). This forces the piston to move down at constant pressure and is the *first part of the power stroke*.

C to D   The fuel supply is cut off at C and the burnt gas expands adiabatically. This forces the piston down and the temperature falls. This is the *second part of the power stroke*.

D to A   The exhaust valve opens at D and releases the exhaust gas, removing an amount of heat (Q_{out}). The pressure and temperature of the gas remaining in the cylinder decrease accordingly.

A to A'   The remainder of the gas is expelled from the cylinder as the piston moves up. This is the *exhaust stroke*.

At A'   The exhaust valve closes, the inlet valve opens and the cycle is repeated.

Like the petrol engine, this is a four-stroke cycle. The main difference between the two engines is that, for the Diesel engine, there is no fuel in the cylinder during compression. Therefore, much higher compression ratios (typically 16:1 for Diesel engines, whereas for petrol engines it is about 10:1) can be achieved, making Diesel engines much more efficient than petrol engines. Again, theoretical efficiency of 65% is not attained because of frictional effects, turbulence, etc., which reduces the efficiency to around 36%.

Diesel engines operate at higher working pressures than petrol engines, which makes them more expensive to produce. They also have a lower power-to-weight ratio.

Real engine cycles

The two p–V indicator diagrams in Figs 15 and 16 are theoretical. In real engines:

- there are no sharp changes, because the valves need time to open and close;
- the expansion and compression strokes are not truly adiabatic, because heat losses occur;
- the maximum temperature is not attained, because of imperfect combustion;
- heating is not achieved at constant volume, because the pistons are always moving.

Power of an engine

The measurement of the power of an engine may be determined or estimated at several points in the transmission of the power from its generation to its application. Three key power indicators are defined in the following ways.

Essential Notes

The compression ratio is the ratio of the volume enclosed in the cylinder at the beginning of the compression stroke to the volume enclosed at the end of the stroke.

The **input power** of an engine is given by

$$P_{\text{input}} - \text{calorific value of fuel (J kg}^{-1}) \times \text{fuel flow rate (kg s}^{-1})$$

Typical calorific values are $44.8\,\text{MJ kg}^{-1}$ for diesel and $48.0\,\text{MJ kg}^{-1}$ for petrol. The fuel flow rate is dependent upon the state of motion of the car, how it is driven, the type of engine, the state of the tyres, etc.

Example 1

A car uses fuel at a rate of $10\,\text{kg h}^{-1}$. Calculate the input power if the fuel has a calorific value of $48\,\text{MJ kg}^{-1}$.

Answer

$$\text{Fuel flow rate} = \frac{10}{3600} = 2.78 \times 10^{-3}\ \text{kg s}^{-1}$$

$$P_{\text{input}} = 48 \times 10^{6} \times 2.78 \times 10^{-3} = 1.3 \times 10^{5}\ \text{W} = 130\ \text{kW}$$

The **indicated power** of an engine is the theoretical capability of the power of an engine, and is given by

$$P_{\text{ind}} = \text{area of } p\!-\!V \text{ loop (J)} \times \text{number of cycles per second (s}^{-1})$$
$$\times \text{number of cylinders}$$

This assumes frictionless motion and so is the maximum theoretical power output of an engine.

Example 2

The $p\!-\!V$ diagram (Fig 17) shows the theoretical cycle for a petrol engine that is running at 1800 rpm. The engine comprises four cylinders. Calculate the indicated power of the engine.

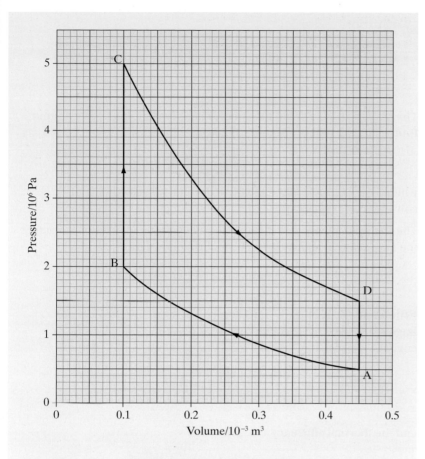

Fig 17
The *p–V* diagram for a theoretical petrol engine

Examiners' Notes

The usual method of determining the work done is by counting the squares under a graph and calculating what the area of one square represents, i.e. the scaling factor. Remember to convert correctly by looking at the units on each axis to ensure that the answer is in joules.

Answer

The work done in one engine cycle is the area enclosed by the loop. Counting small squares gives a value of ~580. Each square represents $(0.1 \times 10^6 \, \text{Pa}) \times (0.01 \times 10^{-3} \, \text{m}^3) = 1 \, \text{J}$ (the scaling factor). So the work done is number of squares × scaling factor = 580 J.

The power cycle occurs once every two revolutions to allow the engine to go through all four of its strokes. Hence the number of cycles per second is given by

$$\text{cycles per second} = \frac{1}{2} \times \frac{1800}{60} = 15 \, \text{s}^{-1}$$

The indicated power is then

$$P_{\text{ind}} = 580 \, \text{J} \times 15 \, \text{s}^{-1} \times 4 = 3.5 \times 10^4 \, \text{W} = 35 \, \text{kW}$$

The **output power** or **brake power** of an engine is a measure of the engine's power without the loss in power caused by the gearbox, alternator, differential, water pump and other auxiliary components. It is the power delivered to the engine's crankshaft (i.e. the engine's flywheel) and is more usually referred to as **brake horsepower** (bhp).

Essential Notes

The idling speed of an engine may be 800 rpm. This means that the angular velocity is given by

$$\omega = \frac{800 \times 2\pi}{60} \text{ rad s}^{-1}$$
$$= 84 \text{ rad s}^{-1}$$

The output or brake power is proportional to the product of the output torque and the number of revolutions per minute.

$$P_{\text{out}} = T\omega$$

where T is the torque in N m and ω is the angular velocity in rad s^{-1}, converted from rpm.

The **frictional power** within an engine can be determined by subtracting the output (brake) power from the indicated power, i.e.

$$P_{\text{friction}} = \text{indicated power} - \text{output (brake) power} = P_{\text{ind}} - P_{\text{out}}$$

Efficiency of an engine

Engine designers make comparisons by looking at the efficiencies of engines. The **mechanical efficiency** η is given by

$$\eta = \frac{\text{output (brake) power}}{\text{indicated power}} = \frac{P_{\text{out}}}{P_{\text{ind}}}$$

and the **thermal efficiency** ε is defined as

$$\varepsilon = \frac{\text{indicated power}}{\text{input power}} = \frac{P_{\text{ind}}}{P_{\text{input}}}$$

The **overall efficiency** of an engine is the product of the mechanical and thermal efficiencies, i.e.

$$\text{overall efficiency} = \eta \times \varepsilon = \frac{\text{output (brake) power}}{\text{input power}} = \frac{P_{\text{out}}}{P_{\text{input}}}$$

Fig 18
The elements of a heat engine

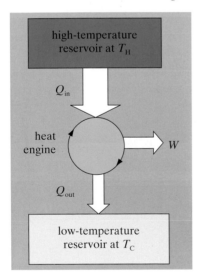

Second law and engines

The essential elements of a heat engine are shown in Fig 18. The arrow on the right indicates the energy extracted as heat from the high-temperature reservoir (the source) that is available to do work. The remainder is delivered to the low-temperature reservoir (the sink).

Essential Notes

The central loop in Fig 18 and its cyclic nature illustrate the function of the working substance, as defined in a p–V diagram.

The purpose of any engine is to transform as much as possible of the extracted energy, Q_{in}, into work. To calculate the net work done by the engine during a cycle, we apply the first law of thermodynamics ($\Delta U = Q - W$) to the working substance. For the idealised heat engine, $\Delta U = 0$ for a complete cycle of the working substance. Hence the work done by the heat engine is given by $W = Q_{\text{in}} - Q_{\text{out}}$. Within the system, there has been no loss of energy.

A measure of the success of a heat engine is through its thermal efficiency, ε, defined as the work the engine does per cycle (the energy we get) divided by the energy it absorbs as heat per cycle (the energy we pay for):

$$\varepsilon = \frac{\text{energy we get}}{\text{energy we pay for}} = \frac{\text{work done in one cycle}}{\text{heat taken in at the higher temperature}}$$

$$\varepsilon = \frac{W}{Q_{in}} = \frac{Q_{in} - Q_{out}}{Q_{in}} = 1 - \frac{Q_{out}}{Q_{in}}$$

The **maximum theoretical efficiency** ε_{max} is given by

$$\varepsilon_{max} = \frac{T_H - T_C}{T_H} = 1 - \frac{T_C}{T_H}$$

where the T_H is the temperature of the hot reservoir (source) and T_C is the temperature of the cold reservoir (sink), both in kelvin.

Because $T_H > T_C$, the heat engine has a theoretical efficiency less than one (i.e. less than 100%). This equation is valid for all reversible engines, irrespective of the particular cycle and the particular working substance.

Ideally, heat should be taken entirely at the single temperature T_H and rejected at the single temperature T_C. However, in real engines, the efficiencies are considerably less than that indicated because the heat is usually taken in over a range of temperatures and rejected over a range of temperatures. Moreover, there are additional losses due to frictional effects, turbulence and other factors.

Designers continually try to improve engine efficiency by reducing the energy Q_{out} that is thrown away during each cycle. However, the perfect engine is not possible. To achieve this, you would need $T_C = 0$ (absolute zero) or $T_H \to \infty$, and both requirements are impossible to attain. The efficiency of a heat engine can only be increased by taking in heat at as high a temperature as possible and rejecting heat at as low a temperature as possible.

Two further observations can be made about heat engines:

- No heat engine converts heat completely into work.

- When a cold body and a hot body are brought into contact, heat always flows from the hot body to the cold body.

These two observations when combined together form a statement that is a version of the so-called **second law of thermodynamics**. This states that it is not possible to convert heat continuously into work without at the same time transferring some heat from a warmer body to a colder body.

In essence, the first law of thermodynamics shows the equivalence of heat and work. The second law shows the conditions under which heat can be converted into work and, in particular, states that heat cannot be converted completely into work.

Real engines, in which the processes that form the engine cycle are not reversible, have much lower efficiencies than those designed. Calculations suggest that the theoretical efficiency of the petrol engine may be as high as 58% but actual efficiencies are nearer to 28%.

Essential Notes

At the completion of the cycle, the engine's working substance is in the same state as it was initially. Therefore, there has been no change in the internal energy ΔU. It follows from the first law of thermodynamics that the work done is equal to the net quantity of heat absorbed, $W = \Delta Q = Q_{in} - Q_{out}$.

Examiners' Notes

When using the theoretical efficiency equation, it is important to change the temperature units into kelvin.

Example

An ideal heat engine operates between the temperatures of 850 K (T_H) and 300 K (T_C). The engine performs 1200 J of work every cycle, taking 0.25 s.

(a) What is the theoretical thermal efficiency of the engine?

(b) What is the average power of the engine?

(c) How much energy is extracted as heat from the high-temperature reservoir every cycle?

(d) How much energy is delivered as heat to the low-temperature reservoir every cycle?

Answer

(a) The theoretical efficiency is

$$\varepsilon = 1 - \frac{T_C}{T_H} = 1 - \frac{300 \, K}{850 \, K} = 0.647$$

$$\varepsilon \approx 65\%$$

(b) Knowing the relation between power and work done gives

$$P = \frac{W}{t} = \frac{1200 \, J}{0.25 \, s} = 4800 \, W = 4.8 \, kW$$

(c) The energy extracted from the high-temperature reservoir is given by the efficiency and the work done per cycle, i.e.

$$\varepsilon = \frac{W}{Q_{in}}$$

so that

$$Q_{in} = \frac{W}{\varepsilon} = \frac{1200 \, J}{0.647} = 1850 \, J$$

(d) Using $W = Q_{in} - Q_{out}$ gives

$$Q_{out} = Q_{in} - W = 1850 \, J - 1200 \, J = 650 \, J$$

Reversed heat engines

Heat engines that work in reverse, i.e. that take heat in at low temperatures and reject heat at higher temperatures, are known as **refrigerators**. A refrigerator is a device that uses work to transfer energy from a low-temperature reservoir to a high-temperature reservoir as it continuously repeats a set of thermodynamic processes. To achieve this practically, some external device (e.g. an electric motor or compressor) has to do work *on* the working substance of the system to transfer energy from say a food storage compartment (low-temperature reservoir) to a room (high-temperature reservoir).

Heat pumps and air conditioners are also refrigerators. For a heat pump that is used to heat a house, the room to be heated is the high-temperature reservoir and heat is transferred to it from the (cooler) outdoors. For the air conditioner, the low-temperature reservoir is the room to be cooled and the high-temperature reservoir is the warmer outdoors. Whatever the system, the diagram in Fig 19 shows the thermodynamic processes involved.

The purpose of the heat pump is to supply heat to the high-temperature reservoir, whereas the purpose of the refrigerator is to remove heat from the cold reservoir. The effectiveness of both heat pumps and refrigerators is determined by the **coefficient of performance (COP)**. This is the ratio of the heat extracted or supplied to the work done by the external agency.

For a refrigerator, the COP is given by

$$\mathrm{COP}_{\mathrm{ref}} = \frac{Q_{\mathrm{out}}}{W} = \frac{Q_{\mathrm{out}}}{Q_{\mathrm{in}} - Q_{\mathrm{out}}} = \frac{Q_{\mathrm{out}}}{Q_{\mathrm{in}}} - 1$$

The maximum theoretical COP for a refrigerator is given by

$$\mathrm{COP}_{\mathrm{max}} = \frac{T_{\mathrm{C}}}{T_{\mathrm{H}} - T_{\mathrm{C}}}$$

For a heat pump, the COP is given by

$$\mathrm{COP}_{\mathrm{hp}} = \frac{Q_{\mathrm{in}}}{W} = \frac{Q_{\mathrm{in}}}{Q_{\mathrm{in}} - Q_{\mathrm{out}}} = 1 - \frac{Q_{\mathrm{in}}}{Q_{\mathrm{out}}}$$

and the maximum theoretical COP for a heat pump is

$$\mathrm{COP}_{\mathrm{max}} = \frac{T_{\mathrm{H}}}{T_{\mathrm{H}} - T_{\mathrm{C}}}$$

Heat pumps provide an extremely low-cost and efficient form of heating, because the heat supplied, Q_{in}, is much greater than the work done by the external device, i.e. $Q_{\mathrm{in}} - Q_{\mathrm{out}}$. The value of the COP is clearly higher the closer the temperatures of the two reservoirs are to each other, when $T_{\mathrm{C}} \sim T_{\mathrm{H}}$. This is why heat pumps work more effectively in temperate climates than in climates where the temperatures vary considerably.

Fig 19
The elements of a refrigerator

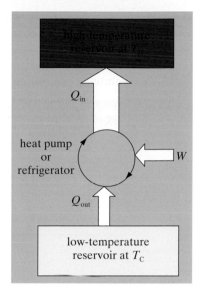

Essential Notes

Values for the coefficient of performance (COP) often greatly exceed 1. Typical household refrigerators and air conditioning units have COPs between 5 and 10. Industrial heat pumps have COP values typically between 10 and 30. A geothermal heat pump with a COP of 4 provides 4 J of heat energy for each unit of work (1 J) put in.

Example

A heat pump working reversibly extracts heat from a river at 8°C and delivers it into a room at 20°C. Determine the coefficient of performance for such a device and explain its significance. How does this compare with a conventional electric fire?

Answer

Converting from °C to kelvin: 8°C = 281 K and 20°C = 293 K.

$$\mathrm{COP}_{\mathrm{hp}} = \frac{Q_{\mathrm{in}}}{W} = \frac{T_{\mathrm{H}}}{T_{\mathrm{H}} - T_{\mathrm{C}}} = \frac{293}{293 - 281}$$

$$\mathrm{COP}_{\mathrm{hp}} = \frac{293}{12} = 24$$

In other words 24 J of heat would be provided with the input of only 1 J of work.

A conventional electric fire can deliver at best 1 J of heat for every 1 J of electrical energy supplied.

Practice exam-style questions

1   Flywheels have recently become the subject of intensive research as power storage devices for use in vehicles and power plants. They are able to store energy very efficiently and then release the energy over a prolonged period of time. A particular portable device has a moment of inertia of 1.2 kg m^2 and can operate safely up to 25 000 revolutions per minute.

   (a)   The flywheel starts at rest and receives an acceleration of 5 rad s^{-2} for 8 minutes. Calculate the final number of revolutions per minute.

_____

_____

_____

_____

3 marks

   (b)   Show that the energy stored in the flywheel at the end of this 8 minute period is 3.5 MJ.

_____

_____

2 marks

   (c)   After reaching its maximum speed, the flywheel is allowed to come to rest. The average power dissipated in overcoming the frictional forces is 15 W. Calculate the time taken before the flywheel is at rest.

_____

_____

_____

2 marks

   (d)   Determine the size of the frictional torque acting on the flywheel.

_____

_____

_____

3 marks

Total Marks: 10

2   A small electric motor is used to start a flywheel. A clutch mechanism is used to connect the motor to the flywheel, when the clutch is said to be engaged.

(a)   Initially the motor is running at 2400 rev min^{-1} and the flywheel is stationary. The motor and clutch system have a moment of inertia of 0.84 kg m^2. Calculate the angular momentum of the motor.

_____

_____

_____

3 marks

(b)   How much rotational kinetic energy does the motor have?

_____

_____

2 marks

(c)   The motor is now connected to the flywheel via the clutch mechanism. The moment of inertia of the flywheel and driveshaft is 1.4 kg m^2. Explain why the speed of the motor decreases as the clutch is engaged.

_____

_____

2 marks

(d)   Calculate the common angular velocity of the motor and flywheel immediately after the clutch is engaged.

_____

_____

_____

_____

3 marks

(e)   Determine the angular impulse on the flywheel as the clutch engages.

_____

_____

_____

_____

2 marks

Total Marks: 12

3   A petrol engine has the following idealised *p–V* diagram (indicator diagram) in which a fixed mass of gas (air) is taken through a cycle involving four processes.

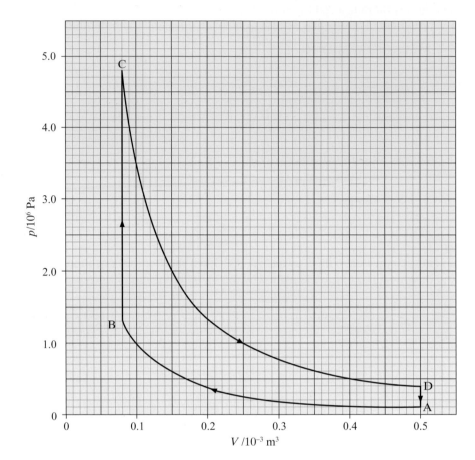

(a)   A → B is an adiabatic compression from an initial temperature of 300 K. Explain what is meant by the *adiabatic compression* of a gas.

_____

_____

1 mark

(b)   B → C is the addition of 850 J of energy at constant volume. Apply the first law of thermodynamics to determine the change in internal energy of the air during this process.

_____

_____

_____

2 marks

(c) Describe the remaining two processes, C → D and D → A within the cycle.

_____

_____

2 marks

(d) Determine the number of moles of air that are taken through the cycle.

_____

_____

_____

_____

_____

3 marks

(e) Determine the work output of the cycle.

_____

_____

_____

3 marks

(f) Real engine cycles differ from the idealised one shown. Give **two** main differences between the theoretical and real cycles.

_____

_____

_____

_____

2 marks

Total Marks: 13

Answers, explanations, hints and tips

Question	Answer		Marks
1 (a)	$\alpha = 5\,\mathrm{rad\,s^{-2}}$ and $t = 8\,\mathrm{min} = 480\,\mathrm{s}$ (remember to convert to seconds)		
	Using the equation		
	$\omega = \omega_0 + \alpha t$	(1)	
	$\omega = 0 + (5)(480) = 2400\,\mathrm{rad\,s^{-1}}$	(1)	
	which converted to $\mathrm{rev\,min^{-1}}$ gives		
	$\omega = 2400 \times \dfrac{60}{2\pi} = 22\,900\,\mathrm{rev\,min^{-1}}$	(1)	3
1 (b)	Using		
	$E_k = \frac{1}{2} I \omega^2 = \frac{1}{2} \times 1.2 \times (2400)^2$	(1)	
	$= 3.5 \times 10^6\,\mathrm{J} = 3.5\,\mathrm{MJ}$	(1)	2
1 (c)	Using		
	$\text{power} = \dfrac{\text{work done}}{\text{time taken}} = \dfrac{E_k}{t}$		
	and rearranging gives		
	$t = \dfrac{E_k}{\text{power}} = \dfrac{3.5 \times 10^6\,\mathrm{J}}{15\,\mathrm{W}}$	(1)	
	$= 2.3 \times 10^5\,\mathrm{s}\,(\approx 64\,\mathrm{h})$	(1)	2
1 (d)	Using		
	$P = T_{\text{average}} \omega_{\text{average}}$	(1)	
	and rearranging gives		
	$T_{\text{average}} = \dfrac{P}{\omega_{\text{average}}} = \dfrac{15\,\mathrm{W}}{(2400\,/\,2)\,\mathrm{rad\,s^{-1}}}$	(1)	
	$= 1.3 \times 10^{-2}\,\mathrm{Nm}$	(1)	3
			Total 10
2 (a)	Angular momentum is		
	$L = I\omega = 0.84 \times \left(\dfrac{2400 \times 2\pi}{60} \right)$	(1)	
	$L = 0.84\,\mathrm{kg\,m^2} \times 251\,\mathrm{rad\,s^{-1}}$	(1)	
	$= 210\,\mathrm{Nms}$ (or $\mathrm{kg\,m^2\,s^{-1}}$)	(1)	3
2 (b)	$E_k = \frac{1}{2} I \omega^2 = \frac{1}{2} \times 0.84 \times (251)^2$	(1)	
	$= 2.6 \times 10^4\,\mathrm{J} = 2.6\,\mathrm{kJ}$	(1)	2
2 (c)	Angular momentum is conserved, i.e. is constant.	(1)	
	Total moment of inertia increases, so ω decreases, hence speed of motor falls.	(1)	2

Question	Answer		Marks
2 (d)	$I = I_{motor} + I_{flywheel}$	(1)	
	$I = 0.84 + 1.4 = 2.20 \text{ kg m}^2$		
	Angular momentum is conserved, hence		
	$L = I\omega_{new}$		
	$\omega_{new} = \dfrac{L}{I} = \dfrac{210 \text{Nms}}{2.20 \text{kgm}^2}$	(1)	
	$= 95 \text{ rad s}^{-1}$	(1)	3
2 (e)	Angular impulse = change in angular momentum		
	Angular momentum of the flywheel before engagement		
	$= I_{flywheel} \times \omega_0 = 0$	(1)	
	Angular momentum of the flywheel after engagement		
	$= I_{flywheel} \times \omega_{new} = 1.4 \times 95 = 130 \text{ Nm s}$		
	Angular impulse $= 130 \text{ N m s}$	(1)	2
		Total 12	
3 (a)	Compression (or decrease in volume) with no heat transfer from the gas.	(1)	1
3 (b)	The first law of thermodynamics states		
	$Q = \Delta U + W$	(1)	
	where ΔU is the increase in internal energy.		
	At constant volume $W = 0$.		
	Hence $Q = \Delta U = 850 \text{ J}$.	(1)	2
3 (c)	C → D is an adiabatic expansion	(1)	
	D → A is a reduction in pressure (cooling) at constant volume	(1)	2
3 (d)	At point A the air in the cylinder has a pressure of $0.1 \times 10^6 \text{ Pa}$ and a volume of $0.5 \times 10^{-3} \text{ m}^3$. The temperature is 300 K.	(1)	
	Using the gas equation,		
	$pV = nRT$		
	and rearranging gives		
	$n = \dfrac{pV}{RT} = \dfrac{0.1 \times 10^6 \times 0.5 \times 10^{-3}}{0.31 \times 300}$	(1)	
	$= 0.020 \text{ mol}$	(1)	3
3 (e)	The work output is the area enclosed by the loop.	(1)	
	Count the number of small squares or use other appropriate method (such as breaking shape up into small trapezia and calculating the area of each).		
	Remember to use the correct scaling factor.		
	Each small square is equivalent to the scaling factor		
	$(0.01 \times 10^{-3}) \text{ m}^3 \times (0.1 \times 10^6) \text{ Pa} = 1 \text{ J}$	(1)	
	Total number of squares is ~350, so work output = 350 J ($\pm$30 J)	(1)	3

Question	Answer	Marks
3 (f)	No sharp corners in real cycles because valves take time to open and close. (1) Expansion and compression strokes are not truly adiabatic in real cycles, as heat losses occur. (1) Real cycles require both induction and exhaust strokes. (1) The maximum temperature is never realised because of imperfect combustion. (1) In real engines, the pistons are always moving so that the heating is not at a constant volume. (1) (any 2)	2
		Total 13

Option Unit 5D Turning Points in Physics

Cathode rays

Fig 1 shows a simplified version of a simple **discharge tube**. This consists of a cold **cathode** (negative electrode), an **anode** (positive electrode) and a gas within an insulating envelope (usually glass). With a sufficiently high potential difference between the cathode and anode, electrons can be liberated from the cathode even at ambient temperatures.

Fig 1
A discharge tube used to demonstrate the existence of cathode rays

The liberated electrons cause a visible discharge in the gas-filled tube

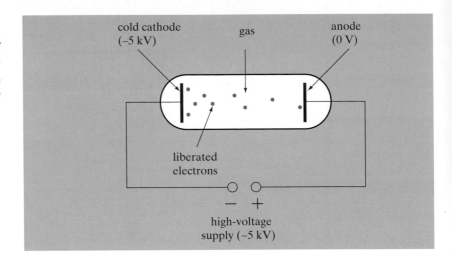

The charged particles are accelerated by an applied potential difference V between the cathode and the anode (typically 5 kV). The nature and properties of these particles were unknown when the effect was discovered in 1876, and the phrase **cathode rays** was used to describe them because they appeared to be emitted from the cathode. By 1897 J. J. Thomson provided significant experimental evidence to show that these rays were in fact electrons.

Essential Notes

The nature of the electron is covered in Unit 1.

Thermionic emission of electrons

When a metal is heated, the free electrons in the metal gain a significant amount of thermal energy. This energy can be sufficient to overcome the attractive electrical force bonding the electrons to the metal, and some electrons break free from the metal surface in a process known as **thermionic emission**. By applying a potential difference between a hot cathode and an anode, an electric field is set up along which the electrons can be accelerated. If the cathode and anode are contained in an evacuated tube and the anode is an annulus containing a small hole at its centre, a narrow beam of electrons can be created. Beyond the anode there is no electric field and the electron beam moves at a constant velocity. Such a device is known as an **electron gun** (Fig 2).

Essential Notes

A cold cathode as in a discharge tube is distinguished from a hot cathode that is heated to induce thermionic emission of electrons.

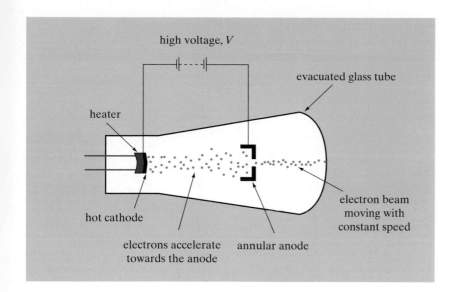

Fig 2
An electron gun uses a heated cathode to liberate electrons, which are accelerated towards the anode. The accelerated electrons are allowed to pass through a small hole in the anode, beyond which they move with a known uniform velocity

The work done on a charged particle when it is accelerated through a potential difference is given by QV, where Q is the charge (in coulombs) and V the potential difference (in volts). For an electron beam, this becomes eV, where e is the **charge of an electron** ($e = 1.60 \times 10^{-19}$ C). This amount of work is equivalent to the kinetic energy gained by the electrons, i.e.

$$\frac{1}{2}mv^2 = eV$$

This allows the definition of an **electron volt** as the kinetic energy of an electron after it has been accelerated through a potential difference of 1 V.

Essential Notes

One electron volt is equivalent to 1.60×10^{-19} J of energy. An accelerating voltage of V volts provides electrons with an energy gain of V electron volts. A more complete description is given in Unit 1.

Example

An electron of mass $m = 9.1 \times 10^{-31}$ kg and charge $e = 1.60 \times 10^{-19}$ C is accelerated through a potential difference of 1.6 kV. Calculate the energy of the electron in electron volts and determine the electron's velocity as it passes the anode.

Answer

Accelerating voltage = 1.6 kV = 1600 V. Hence the electron gains 1600 eV. Using

$$\frac{1}{2}mv^2 = eV$$

and rearranging gives

$$v = \sqrt{\frac{2eV}{m}} = \sqrt{\frac{2 \times 1.60 \times 10^{-19} \times 1600}{9.1 \times 10^{-31}}} = 2.4 \times 10^7 \text{ m s}^{-1}$$

or about 8% of the speed of light, i.e. it is travelling at relativistic speed.

Determination of the specific charge of an electron, e/m, by any one method

J. J. Thomson discovered the electron in 1897 using a modified version of the discharge tube (Fig 1). A modern **cathode ray tube** used to determine e/m is shown in Fig 3.

Fig 3

Modern cathode ray tube used to determine the ratio of e/m. The direction of the magnetic field (B) is into the paper (denoted by ×). The electric field (E) is shown by the red arrows

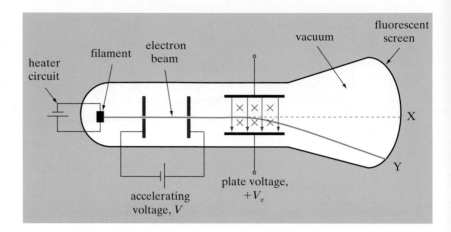

Electrons, emitted thermionically from a metal filament, are accelerated towards the anode in a fine beam and pass through it. Beyond the anode, the electrons are subjected to both an electric field and a magnetic field at right angles to each other and to the electron beam. The force on the electrons in the uniform electric field, E, is given by

$$F = eE = \frac{eV_e}{d}$$

where V_e is the potential difference between the deflecting plates and d is their known separation. The path traced out by a beam in an electric field alone can be shown to be parabolic. However, the magnetic field, B, provides another force on the electrons given by

$$F = Bev$$

where v is the electron's velocity. This force is perpendicular to both the field direction and the velocity. The path taken by the electrons in a magnetic field alone is along a circular arc; the magnetic field does not change the kinetic energy of the electrons.

When both fields (E and B) are zero, the electron beam reaches the fluorescent screen at X. With just the magnetic field turned on, the electrons are deflected to a new position Y. Thomson then switched on the electric field and adjusted this field until the beam was once again at position X. Under these circumstances, the forces exerted by each field must be equal, and therefore

$$eE = Bev$$

so that

$$v = \frac{E}{B}$$

Electric and magnetic fields in this configuration are known as **crossed fields** and can be used to obtain charged particles of a particular velocity from a beam containing particles with a range of velocities.

The velocity of the electrons leaving the anode is given from their kinetic energy,

$$eV = \frac{1}{2}mv^2$$

where V is the potential difference between anode and cathode. From this expression, the ratio e/m can be obtained as

$$\frac{e}{m} = \frac{v^2}{2V}$$

Substituting for v gives

$$\frac{e}{m} = \frac{E^2}{2VB^2}$$

A more elegant method uses a beam of electrons emitted into a chamber with a velocity v using an electron gun, known as a **fine beam tube** (Fig 4). The chamber is filled with hydrogen gas at low pressure. As the electrons pass through the chamber, they collide with hydrogen atoms, which absorb some energy from the bombarding electrons and enter an excited state. Almost immediately the hydrogen atoms de-excite and fall back into their ground state by emitting photons in the visible region of the spectrum. The path of the electron beam is then clearly visible within the chamber.

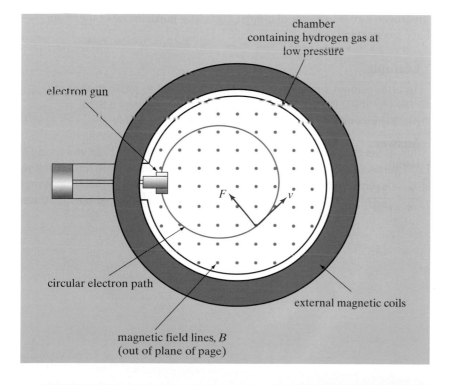

Fig 4
A fine beam tube, showing the experimental arrangement to measure e/m. The magnetic field lines are out of and perpendicular to the page and are represented by the array of dots

Fig 6

(a) With no electric field applied, a spherical oil drop may fall through a viscous fluid at a constant (terminal) velocity

(b) With an electric field applied, the oil drop may be held stationary

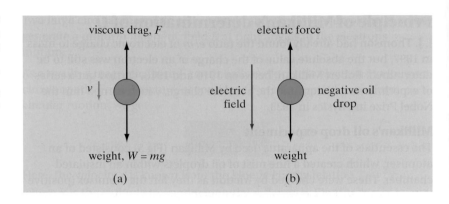

viscous drag, F electric force

electric field   negative oil drop

v

weight, $W = mg$ weight

(a)   (b)

Examiners' Notes

The expression for Stokes' law will be provided. You need to understand that F is a viscous force.

Millikan used **Stokes' law** to determine the viscous force on the drop. For a spherical object moving with a velocity v, the viscous force F is given by

$$F = 6\pi\eta r v$$

where η is the **viscosity** of the fluid, and r is the radius of the drop. With the forces in balance,

$$mg = 6\pi\eta r v$$

Since mass = volume × density, this can be rewritten as

$$\frac{4}{3}\pi r^3 \rho g = 6\pi\eta r v$$

where ρ is the density of the fluid. This can be rearranged to give

$$r = \left(\frac{9\eta v}{2\rho g}\right)^{1/2}$$

Millikan found both η and ρ in separate experiments and hence he could determine r.

Essential Notes

The electric force is proportional to the potential difference across the plates. This causes the oil droplet to decelerate.

By applying a voltage between the plates P_1 and P_2 (Fig 5), a uniform electric field was established and an electric force was exerted on the oil drop. If the oil drop charge Q is negative, then the electric force on the drop acts upwards. Varying the potential allowed Millikan to control the motion of the drop until it was held stationary (Fig 6b).

The viscous force had disappeared and the two forces involved were in balance, i.e. the weight and the electric force:

$$mg = \frac{QV}{d}$$

where d is the distance between the plates. Once again, this can be written as

$$\frac{4}{3}\pi r^3 \rho g = \frac{QV}{d}$$

or

$$Q = \frac{4\pi r^3 \rho g\, d}{3V}$$

All the quantities on the right-hand side of the equation are known, and so Millikan was able to determine the charge Q.

Example

A charged oil droplet falls between two plates 1.2 mm apart. When the potential difference between the two plates is zero, the droplet falls vertically at a steady speed of $6.2 \times 10^{-5}\,\mathrm{m\,s^{-1}}$.

(a) If the density of the oil droplet is $960\,\mathrm{kg\,m^{-3}}$ and the viscosity of air is $1.8 \times 10^{-5}\,\mathrm{N\,s\,m^{-2}}$, find the radius of the droplet.

(b) Find the mass of the droplet.

(c) If the droplet is stationary when a potential difference of 60 V is applied, find the magnitude of the charge on the droplet and discuss its significance.

Answer

(a) Substituting into the expression

$$r = \left(\frac{9\eta v}{2\rho g}\right)^{1/2} = \sqrt{\frac{9 \times 1.8 \times 10^{-5} \times 6.2 \times 10^{-5}}{2 \times 960 \times 9.81}} = 7.3 \times 10^{-7}\,\mathrm{m}$$

(b) Using mass = volume × density,

$$m = \frac{4\pi r^3 \rho}{3} = \frac{4\pi \times (7.3 \times 10^{-7})^3 \times 960}{3} = 1.60 \times 10^{-15}\,\mathrm{kg}$$

(c) Using electric force (Qv/d) = weight (mg),

$$Q = \frac{mgd}{V} = \frac{1.60 \times 10^{-15} \times 9.81 \times 1.2 \times 10^{-3}}{60} = 3.1 \times 10^{-19}\,\mathrm{C}$$

This is approximately equal in size to a charge of $2e$.

Significance of Millikan's results

By repeating the experiment many times, Millikan found that the values of Q were always given by

$$Q = ne \quad \text{for } n = \pm 1,\ \pm 2,\ \pm 3,\ \ldots$$

in which e turned out to be the fundamental constant that we now know is the elementary charge of an electron. This result was really significant. Millikan concluded that charge can never exist in smaller quantities than this and it was convincing proof that charge is quantised. He also proposed, correctly, that the charge of the electron was $1.60 \times 10^{-19}\,\mathrm{C}$.

Wave–particle duality

Newton's corpuscular theory of light

In 1672 Sir Isaac Newton published his theory of colour, in which he suggested that light was composed of tiny particles called **corpuscles**. At that time, light was known both to reflect and to refract (but not known to diffract), and using his own laws of motion Newton was able to demonstrate his **corpuscular theory**. Newton explained reflection simply as a force that pushed away the corpuscles from the surface (Fig 7a). The velocity component parallel to the surface remained unchanged. He said that the effects of refraction were due to the fact that the corpuscles travelled faster (not slower) in a denser medium. The component of the velocity (or momentum) parallel to the surface was unchanged (Fig 7b), whereas the component of the velocity (or momentum) normal to the surface increased due to the force of attraction.

Fig 7
Newton's explanation of reflection and refraction

(a) Reflection: $v \sin \theta_1 = v \sin \theta_2$,
so $\theta_1 = \theta_2$

(b) Refraction:
$v_1 \sin \theta_1 = v_2 \sin \theta_2$,
so $\dfrac{\sin \theta_1}{\sin \theta_2} = \dfrac{v_2}{v_1}$

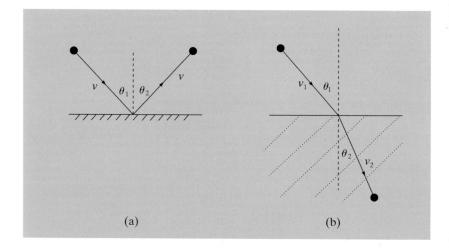

(a) (b)

However, in 1678, an alternative model for light was proposed by the Dutch physicist Christian Huygens, based on waves. **Huygens' wave theory** was based on a geometrical construction that allowed a given future **wavefront** to be located if its present position was known. The construction is based on what is now called **Huygens' principle** (Fig 8). This states that all points on a wavefront serve as point sources of spherical secondary wavelets that spread out in the forward direction at the speed of the wave. The new wavefront position is the surface that is tangential to all of these secondary wavelets.

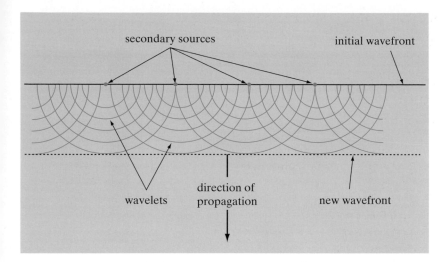

Fig 8
Huygens' geometrical construction, now known as Huygens' principle

The great advantage of Huygens' wave theory was that it could account for the laws of reflection and refraction (Fig 9), and gave physical meaning to the index of refraction (n) by showing that light slowed down when entering a denser medium, rather than speeding up as in the corpuscular theory. More importantly, Huygens predicted that light should undergo **diffraction** around small objects and that two **coherent light sources** should show **interference**.

Essential Notes

Coherent light sources are sources of light that have a constant phase difference. In Huygens' time, obtaining two independent sources of light that were coherent was difficult to achieve experimentally.

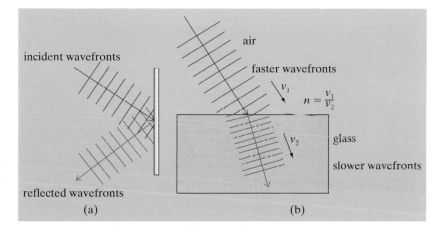

(a)   (b)

Fig 9
Huygens' principle applied to
(a) light reflection
(b) light refraction

Essential Notes

The refractive index of a material, n, is the ratio of the velocity of light in air (strictly a vacuum), v_1, to the velocity of light in the material, v_2:

$$n = \frac{v_1}{v_2}$$

Unfortunately, Huygens' wave theory was not readily accepted. Newton's reputation was immense, and the majority of scientists accepted the corpuscular model. The apparent conflict between the two models continued for over 100 years until in 1801 Thomas Young experimentally proved that light is a wave, contrary to what other scientists then thought.

Significance of Young's double slits experiment

Young demonstrated that light undergoes interference just like water waves and sound waves. The problem of using two coherent light sources was solved by using just one light source in front of two narrow slits in what became known as Young's double slits experiment (Fig 10).

Fig 10
Young's double slits experiment to show diffraction and interference

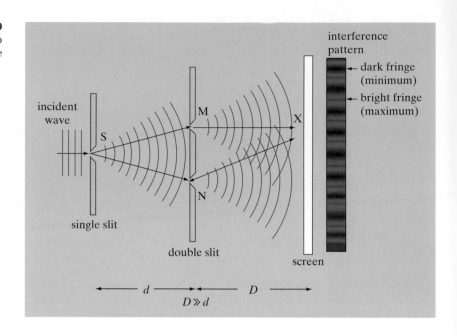

Essential Notes

The difficulties of achieving coherent light sources and a double slits interference pattern is covered in Unit 2, as is the diffraction of light by a single slit.

Incident monochromatic light (light of a single wavelength) is diffracted by the first slit S, which then acts as a point source of light that emits semicircular wavefronts. These wavefronts are diffracted by the double slits, M and N, which act as two point sources of light. These new wavefronts then overlap and interfere, at points such as X, producing a pattern of maxima (bright fringes) and minima (dark fringes) on the viewing screen, called an interference pattern.

In 1850 Foucault measured the velocity of light in air and in water. He found that the velocity of light in water was *lower* than that in air. Since it was shown that light was refracted towards the normal in moving from air to water, Foucault's result confirmed the prediction of the wave nature of light. This was conclusive proof that light both diffracts and interferes, and a final acceptance that Huygens was correct and Newton was wrong.

Electromagnetic waves

Essential Notes

See Unit 2 for a discussion of the polarisation of transverse waves. Radio waves from a transmitter are polarised and hence the aerial of a radio receiver needs to be aligned in the same plane as the transmitter.

Light waves, like sound waves, were thought to behave as a longitudinal wave. In 1810 the **polarisation** of light was observed for the first time, and this could not be explained in terms of longitudinal waves. By 1817 Thomas Young had suggested that light was a transverse wave, which explained its polarisation. James Clerk Maxwell's achievement was to show theoretically that a beam of light is a transverse wave consisting of vibrating electric and magnetic fields travelling at right angles to each other and to the direction of travel (Fig 11). This is known as an **electromagnetic wave**. The study of visible light is simply the study of a small region of the complete electromagnetic spectrum.

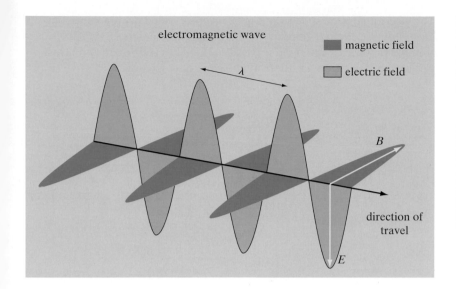

electromagnetic wave

■ magnetic field

□ electric field

λ

B

direction of travel

E

Fig 11
Electromagnetic waves are not vibrations of particles, but oscillating electric and magnetic fields. These fields are perpendicular to each other and to the direction of wave travel

Examiners' Notes

You should be able to state and draw a sketch to show that the direction of propagation is perpendicular to both the electric and the magnetic waves.

During the mid-1800s (in Maxwell's time), the visible, infrared and ultraviolet forms of light were the only electromagnetic waves known. In 1887 Heinrich Hertz confirmed Maxwell's predictions by his discovery of 'Hertzian waves', now known as radio waves, and verified that they travelled at the same speed as visible light. Hertz also showed that these radio waves could be reflected, refracted and diffracted, and displayed both interference and polarisation phenomena. In fact, all electromagnetic waves show these properties.

Maxwell showed that an electromagnetic wave comprised sinusoidally oscillating electric, E, and magnetic, M, fields, in phase with one another. The ratio of the field strengths at any point in time was given by

$$\frac{E}{M} = c$$

where c is the wave speed, i.e. the speed of light. Maxwell furthermore showed that this ratio is also given by the expression

$$c = \frac{1}{\sqrt{\mu_0 \epsilon_0}}$$

where μ_0 is the **permeability** of free space and ε_0 is the **permittivity** of free space. Both of these are constants, with $\mu_0 = 4\pi \times 10^{-7}\,\text{H m}^{-1}$ and $\varepsilon_0 = 8.85 \times 10^{-12}\,\text{F m}^{-1}$. The permittivity of free space relates to the electric field strength due to a charged particle, and the permeability of free space relates to the magnetic flux density due to a current-carrying wire.

Essential Notes

Permittivity is associated with capacitance (see Unit 4) and has units of farad per metre (F m^{-1}); $1\,\text{F} = 1\,\text{A s V}^{-1}$.

Permeability is associated with inductance and has units of henry per metre (H m^{-1}); $1\,\text{H} = 1\,\text{V s A}^{-1}$.

The units of c are therefore

$$c = \frac{1}{\sqrt{\dfrac{\text{V} \times \text{s}}{\text{A} \times \text{m}} \times \dfrac{\text{A} \times \text{s}}{\text{V} \times \text{m}}}} = \text{m s}^{-1}$$

as expected.

The electromagnetic spectrum (Fig 12) is open ended, with no upper or lower limits. Certain regions are identified by familiar labels, such as X-rays, microwaves and radio waves, but there is a continuous progression through the spectrum, i.e. there are no gaps.

Fig 12
The electromagnetic spectrum, showing the position of the visible region

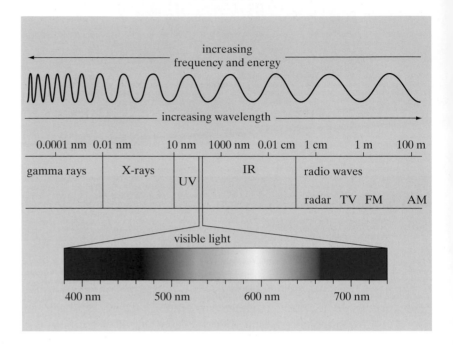

Essential Notes

All electromagnetic waves travel at the same speed in a vacuum, that is $2.998 \times 10^8 \, \text{m s}^{-1}$ (the Data and Formulae Booklet gives $3.00 \times 10^8 \, \text{m s}^{-1}$ to three significant figures). The properties of the wave and the way it interacts with matter depend upon its wavelength, λ.

All electromagnetic waves, no matter where they are along the spectrum, travel through free space with the same speed. Maxwell in 1862 calculated this to be $2.998 \times 10^8 \, \text{m s}^{-1}$. The speed of light in other media is lower than this.

Properties of electromagnetic radiation

It is convenient to divide the spectrum into seven major regions that show the main characteristics of the radiation (Table 1). This extends from long to short wavelengths, and includes radio waves, microwaves, infrared radiation, visible light, ultraviolet radiation, X-rays and gamma radiation. Electromagnetic waves are classified by their origin rather than by wavelength alone. An X-ray and a gamma ray may have exactly the same wavelength, and therefore properties, but have arisen from different physical origins.

Region	Wavelength range/m	Origin	Use
radio waves	10^{-1} to 10^{6}	oscillations in electric fields	• radio transmissions
microwaves	10^{-3} to 10^{-1}	molecular interactions	• radar • TV and mobile transmissions • cookery
infrared (IR)	10^{-7} to 10^{-3}		• heat detectors • night vision cameras • TV and games remote controls • optical fibres
visible	4×10^{-7} to 7×10^{-7} (400–700 nm)	energy transitions within atoms	• optical fibres • (vision) range of eye sensitivity
ultraviolet (UV)	10^{-9} to 10^{-7}		• fluorescence • security markings
X-rays	10^{-13} to 10^{-8}		• medical and dental imaging and diagnosis • airport security scanners • medical cancer treatment
gamma rays	10^{-16} to 10^{-10}	nuclear de-excitation	• irradiation and sterilisation of food and equipment • medical cancer treatment

Table 1
The seven major areas associated with the electromagnetic spectrum, their origin and their uses

The discovery of photoelectricity

The photoelectric effect

If a direct beam of light or other electromagnetic waves of high enough frequency is directed onto a clean metal surface, the waves interact with free electrons close to the surface and cause them to vibrate. If these electrons absorb sufficient energy, they are liberated or ejected from the surface as **photoelectrons** almost instantaneously, in an effect known as the **photoelectric effect** (Fig 13). The effect was first discovered in 1887 by Heinrich Hertz using a spark gap detector. Hertz noticed that strong sparks were produced when the detector was illuminated with ultraviolet light, but did not attempt to explain the effect. However, the emission of photoelectrons is readily demonstrated using a gold-leaf electroscope, a device that is very sensitive to charge. When the electroscope is positively charged, the thin gold leaf rises as it is repelled from the metal stem (like charges repel) and remains in that position. When ultraviolet radiation from a UV lamp is directed at the surface of a clean zinc plate placed on the cap of the electroscope, the gold leaf remains in position. However, when the electroscope is negatively charged, the effect of ultraviolet radiation allows the gold leaf to fall gradually. The leaf falls because photoelectrons are emitted from the zinc surface. The photoelectric effect

Essential Notes

A spark gap detector is a device that has a small gap between two electrodes, across which a spark passes on ionisation of the air in the gap.

was investigated thoroughly by Millikan and others. An explanation of Millikan's results for sodium is given later, in Fig 15.

Fig 13
The basic process of emission of photoelectrons from a metal surface

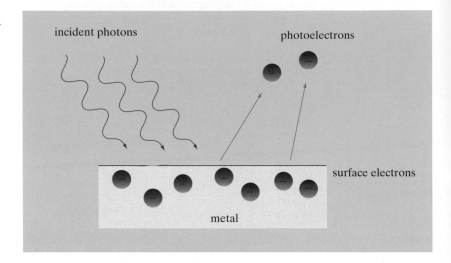

The photoelectrons are emitted with a range of kinetic energies, from zero to some maximum value. This value of maximum kinetic energy increases with the frequency of the radiation but is unaffected by the intensity of the radiation. For example, a faint ultraviolet glow would cause the emission of more energetic electrons than an intense red laser beam.

If in Fig 14 the potential difference, V, is changed so that the collector (C) is slightly negative with respect to the target (T), then the maximum kinetic energy of the photoelectrons can be found. The value of V is varied until the most energetic photoelectrons do not reach C (i.e. the current falls to zero), and this value is called the **stopping potential**, V_s. The maximum kinetic energy of the photoelectrons is then given by

$$\frac{1}{2}mv^2_{max} = eV_S$$

where e is the charge of an electron (1.60×10^{-19} C). Measurements have shown that, for a given frequency, the maximum kinetic energy does not depend on the intensity of the light source (i.e. whether it is bright or feeble).

A second experiment involves varying the frequency of the incident light, by use of suitable filters, and measuring the stopping potential. For a given metal, no photoelectrons are emitted if the radiation has a frequency below a certain value called the **threshold frequency** or cut-off frequency, f_0, again independent of the intensity of the incident beam. For example, if the metal zinc is illuminated with visible light, no electrons are emitted; however, if ultraviolet light is used, there is a significant release of photoelectrons. Also, photoelectric emission begins immediately the radiation falls on the surface, no matter how dim, provided the frequency exceeds f_0. All of these effects were a puzzle for classical physics.

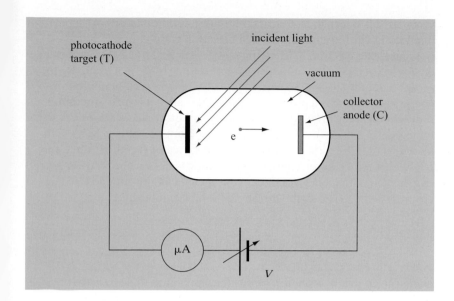

Fig 14
Using a vacuum photocell to
evaluate the stopping potential

Classical physics

According to the wave theory of light, for a particular frequency, the energy
carried is proportional to the intensity (amplitude) of the wave, and this
would be spread evenly over the wavefront. Each free electron at the
surface of the exposed metal would gain a small amount of energy from
each incoming wave, and after a period of time each electron would have
gained sufficient energy to escape from the metal. Therefore, according to
wave theory, it should take a finite time for the emission of photoelectrons
to begin. Instantaneous emission cannot be explained.

Light of lower frequency, and hence less energy, incident on the metal
surface would mean that it would take longer for the electrons to gain
sufficient energy to be liberated. The notion of a threshold frequency
cannot be explained.

The greater the intensity of the incident wave, the larger the amount of
energy transferred to the electron, so the kinetic energy should also
increase with intensity. The dependence of the kinetic energy on frequency
cannot be explained.

Einstein's explanation of the photoelectric effect

To explain these observations, in 1905 Einstein used the ideas put forward
by Max Planck that electromagnetic waves can only be emitted in discrete
packets or quanta. Planck proposed that the energy carried by one of these
quanta was related to its wavelength via the expression

$$E = hf = \frac{hc}{\lambda}$$

where c is the speed of light in free space and h is known as the Planck
constant (6.63×10^{-34} J s). Einstein called these individual packets
'photons', and he visualised the photoelectric effect as an interaction
between one individual photon and one electron, with the photon giving
all of its energy to that electron.

Examiners' Notes

In a photon–electron
interaction, one electron
absorbs one photon, and
consequently the photon
disappears.

When light is directed onto a metal surface it is bombarded with a beam of photons. During an interaction, an electron gains energy $E = hf$.
In order for an electron to leave the metal surface, it must have sufficient energy to overcome the attractive electrical force holding it there. To just escape from the surface, an electron must acquire a certain minimum energy, Φ, where Φ is a property of the metal called its **work function**. If the energy transferred to an electron by a photon exceeds the work function, i.e. $hf > \Phi$, the electron can escape from the surface as a photoelectron. If $hf < \Phi$ the electron vibrates, heating up the surface and releasing another, lower-energy, photon; it does not escape. The existence of a threshold frequency is a direct consequence of this interaction. For electrons to be released, the threshold frequency must be

$$f_0 = \frac{\Phi}{h}$$

The kinetic energy that an electron has when it leaves the surface is the energy given to it by a photon minus the work it has to do to leave the surface. Some electrons come from slightly below the surface, make collisions, and have to do more work than Φ to escape. Hence the kinetic energy of the emitted electrons ranges from zero up to some maximum value. The minimum energy that a surface electron can lose is the work function. Hence the maximum kinetic energy is given by

$$\frac{1}{2}mv_{\text{max}}^2 = hf - \Phi$$

and the maximum kinetic energy may be determined from the stopping potential. Fig 15 gives an explanation of Millikan's results for sodium metal, which could not be explained by classical physics.

Essential Notes

Further details of the photoelectric effect can be found in Unit 1.

Fig 15
Millikan's experimental results for sodium metal. The gradient is equal to the Planck constant, h. The intercept on the vertical axis is the work function, Φ, of sodium (see explanation in text). The intercept on the horizontal axis is the threshold frequency, f_0, for sodium

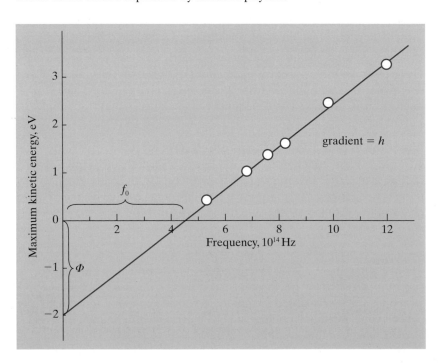

Einstein's explanation was a compelling argument for the existence of photons and that light is in fact quantised as photons.

Wave–particle duality

While the photoelectric effect showed light quantised as photons, diffraction and interference of light showed light as waves. Neither the particle picture nor the wave picture describe fully what light is – they are simply two different ways to explain how light behaves. These are examples of what is referred to as **wave–particle duality**. Despite the terminology, there is no contradiction here – the wave and particle theories are complementary to each other. In 1916 Einstein extended his concept of light quanta (photons) still further by proposing that a photon also has linear momentum, p. For a photon with energy hf, the size of this momentum is

$$p = \frac{hf}{c} = \frac{h}{\lambda}$$

When a photon interacts with matter, both energy and momentum are transferred, and this was experimentally vindicated by the American physicist Arthur Holly Compton in 1923 (the so-called Compton effect).

The idea that this concept of wave–particle duality could be applied to electrons and other particles was put forward by Louis de Broglie in 1924. He suggested that *all* particles should be thought of as matter waves, and he assigned a wavelength λ to a particle with momentum p (mass m and speed v) via the relation

$$\lambda = \frac{h}{p} = \frac{h}{mv}$$

where h is the Planck constant. The wavelength of the particle is called the **de Broglie wavelength**.

De Broglie's prediction of the existence of matter waves was verified experimentally in 1927 through electron diffraction studies by Davisson and Germer and a year later by the studies of George Thomson (the son of J. J. Thomson)

Electron diffraction

In the Davisson and Germer experiment, a fine beam of accelerated electrons within a vacuum tube were directed towards a target material made up of a powder of tiny graphite crystals. Diffraction occurs only when the electron interacts with an object of about the same size as its de Broglie wavelength, λ. This is related to the accelerating voltage by the expression

$$\lambda = \frac{h}{\sqrt{2meV}}$$

where e is the charge of the electron, V is the accelerating voltage and m is the electron mass. This is the rest mass for non-relativistic speeds, at low V, whereas the relativistic mass (see page 218) should be used for high values of V.

Essential Notes

Wave–particle duality is also covered in Unit 1.

Essential Notes

The formulation of this expression and the application of electron diffraction to determine nuclear diameters is covered in the core of Unit 5, page 21.

Examiners' Notes

This expression is given in the Data and Formulae Booklet.

Essential Notes

There are at least as many experiments as those that support the particle nature of matter.

By increasing *V*, the speed of the electrons could be increased and hence their wavelength decreased. The results of the experiment produced circular diffraction patterns on a photographic film screen (Fig 16). Small accelerating voltages produced slower electron beams and widely spaced concentric rings. Larger voltages produced higher electron speeds and more tightly spaced rings, as predicted by the de Broglie formula. This was convincing evidence for the wave nature of matter and the most obvious confirmation of the developing theory of quantum mechanics.

Fig 16
Electron diffraction pattern showing a series of concentric rings

Essential Notes

Similar interference patterns have been demonstrated with protons, neutrons and various atoms, including molecules of iodine that are 500 000 times more massive than electrons and far more complex.

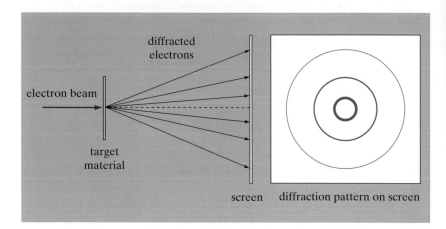

Because of the dependence of the diffraction pattern on the distances between atoms, electron diffraction is now also an important tool for determining the structure of crystals, free molecules, gases and liquids. Both low-energy and high-energy electron diffraction are used routinely.

Example

A beam of electrons is accelerated across an electric potential of 5 kV. The electrons are incident upon a powder of tiny crystals in a target. Calculate the de Broglie wavelength of the electrons and comment on its value. Take the mass of an electron $m = 9.11 \times 10^{-31}$ kg and its charge $e = 1.60 \times 10^{-19}$ C.

Answer

First find *p* using

$$p = \sqrt{2meV} = \sqrt{2 \times 9.11 \times 10^{-31} \times 1.60 \times 10^{-19} \times 5000} = 3.82 \times 10^{-23} \text{ kg m s}^{-1}$$

Using the de Broglie condition,

$$\lambda = \frac{h}{p} = \frac{6.63 \times 10^{-34}}{3.82 \times 10^{-23}} = 1.7 \times 10^{-11} \text{ m}$$

This is of the same order of magnitude as the atomic spacing within the target crystals.

Electron microscopes

An **electron microscope** uses a beam of electrons to 'illuminate' a specimen and create a highly magnified image. Electron microscopes have a much greater resolving power, and hence greater useful magnification, than optical microscopes because the wavelength of the electron (its de Broglie wavelength) is much smaller than that of a photon of visible light. This may be of the order of 10^3–10^4 times smaller than optical wavelengths. The anode potential V needed to produce a wavelength of the order of the size of an atom can be found by rearranging the expression for λ given in the previous section (page 209), to obtain

$$V = \frac{h^2}{2\lambda^2 me}$$

For an operating voltage of 3000 V, the wavelength is about 2×10^{-11} m, and hence comparable to the size of atoms.

Electron microscopes allow the detailed examination of tiny structures that would otherwise be blurred by diffraction if viewed with an optical microscope. Electron microscopes use thermionic emission from a filament, usually tungsten, to generate a beam of electrons. These electrons are then accelerated by an electrical potential difference and focused by electromagnetic lenses onto a sample. These lenses have a similar effect to the glass lenses in optical microscopes. There are two main types of electron microscope.

A **transmission electron microscope** (**TEM**) passes a beam of electrons through an ultra-thin specimen (Fig 17). The monochromatic electron beam is generated by an electron gun. The initial beam of electrons is focused by a magnetic condenser lens onto the specimen. The electrons are scattered at different angles depending on the arrangement of atoms within the specimen. The objective aperture, situated below the specimen, constrains the scattering angle to no more than 0.5 degrees (typically), knocking out electrons far from the optical axis. An image (Fig 18) is formed from the interaction of the electrons that are transmitted through the specimen, and these are focused by an intermediate lens and projector lens system onto a fluorescent screen or photographic film.

Darker areas of the image represent those areas within the specimen through which fewer electrons were transmitted because it is thicker or denser. Conversely, lighter areas of the image represent those areas of the specimen where electrons were transmitted because it is thinner or less dense.

A monochromatic beam of electrons is essential to avoid aberrations due to some electrons travelling faster than others within the beam. Similarly, a thin specimen is also crucial, as this affects the resolution of the microscope: thicker specimens have a greater impact on slowing the electrons, increasing their wavelength and hence reducing the details that can be observed.

TEMs form a major analysis technique in physical and materials science, and have found numerous applications in many areas of biological science, including cancer research and virology.

Fig 17
A schematic diagram of a transmission electron microscope (TEM), showing the arrangement of magnetic lenses used to produce an image

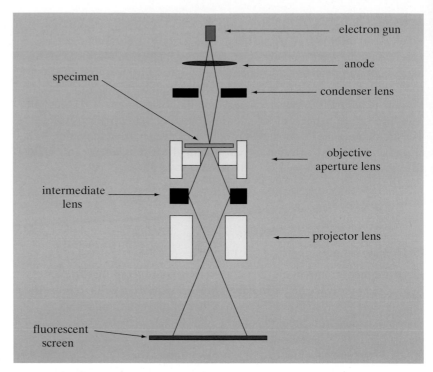

electron gun

anode

specimen

condenser lens

objective aperture lens

intermediate lens

projector lens

fluorescent screen

Fig 18
A TEM image of an ordered array of single-crystal gold (Au) nanoclusters

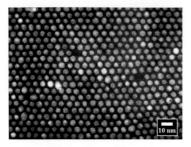

10 nm

Fig 19
An STM image of individual silicon atoms arranged on a face of a crystal

Essential Notes

An STM operates in two modes:
(i) in constant-height mode, the current changes as the gap varies, and the variation in current with time is used to map the surface;
(ii) in constant-current mode, the gap width is measured as the current is kept constant, by varying the height of the tip above the specimen.

A **scanning tunnelling microscope (STM)** is based on the concept of quantum tunnelling and was developed in 1981 by Binnig and Rohrer. A very fine conducting tip is brought very near to a metallic or semiconducting surface. Electrons tunnel through the vacuum between the tip and the surface, producing an electrical signal. By scanning across a specimen, variations in current are translated into an image (Fig 19). The smaller the distance between the tip and the specimen surface, the greater the current.

STMs can be used not only in ultra-high-vacuum conditions but also in air and over a range of temperatures from near absolute zero to a few hundred degrees Celsius. STMs have spatial resolutions of 0.1 nm and depth resolutions of 0.01 nm (smaller than any atom). The versatility of the microscope earned the inventors the Nobel Prize in Physics in 1986.

D.4.3 Special relativity

The Michelson–Morley experiment

According to Newton's laws, any object that is in motion is moving relative to some fixed background. In the late 19th century, most scientists believed that light or any electromagnetic wave required a medium in which to travel. They supposed that such a wave could not travel through a vacuum. This fixed background or medium was called the **ether** (or aether) and all objects and radiation therefore moved with an 'absolute motion' relative to this ether. In 1887 Michelson and Morley devised an experiment to determine the absolute speed of the Earth through the ether using a modified interferometer originally designed by Michelson.

Michelson and Morley expected the motion of the Earth to affect the speed of light as it moved through the ether. A measurement of the speed of light, c, measured on Earth, which is moving with a speed v parallel to the light, would be either $(c + v)$ or $(c - v)$. Michelson and Morley were to measure the speed of light parallel and perpendicular to the motion of the Earth, and hence to determine the velocity v, the absolute speed of the Earth.

Principle of the Michelson–Morley interferometer

The **Michelson–Morley interferometer** in its simplest form (Fig 20) comprises two plane mirrors and a partial plane mirror, called a beam splitter. The partial mirror allows 50% of the light to be transmitted and the remainder to be reflected, making two identical but separate beams of light.

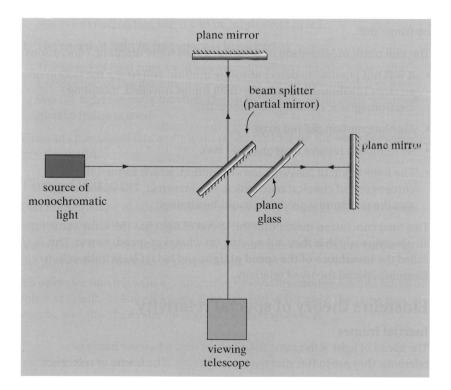

Essential Notes

Scientists assumed that the Sun was approximately stationary in the ether; hence the speed measured in the experiment would be the Earth's speed about the Sun.

Essential Notes

An interferometer is a device used to measure wavelengths by the production of interference fringes.

Fig 20
A simplified version of the Michelson–Morley interferometer

Essential Notes

In the original interferometer, the light beam went back and forwards several times before being split to give an effective path length of 10 m for each arm. The whole system was mounted on a heavy slab suspended on a pool of mercury for ease of rotation.

where c is the speed of light ($3.00 \times 10^8\,\mathrm{m\,s^{-1}}$) and the term $(1 - v^2/c^2)^{-1/2}$ is called the **Lorentz factor** or relativity factor, γ (see Fig 22), which appears quite frequently in the theory of relativity. Often the above expression is more simply written as

$$t = \gamma t_0$$

Measurements of a proper time interval from any other inertial frame of reference are always longer (greater). This is called **time dilation**.

Fig 22
The Lorentz factor or relativity factor, γ, shows the dramatic effects of relativity when v approaches the speed of light

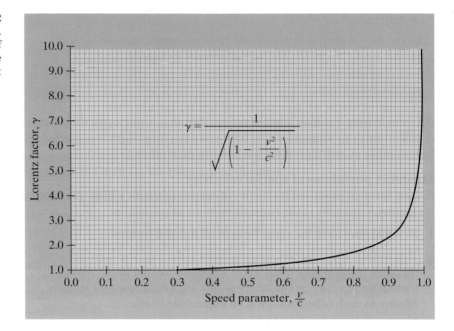

Example

An observer A, travelling on a train at a constant speed of 80% of the speed of light, switches on a torch for exactly 3 s as the train passes through a station. Another person B, standing stationary on the platform, records the same event but for a longer time. What time does B record?

Answer

In this example, the proper time (t_0) is 3 s as measured by A, so A is the stationary observer in this case. Observer B is moving at $0.8c$ relative to the event, so B's time is given by

$$t = t_0 \left(1 - \frac{v^2}{c^2}\right)^{-1/2} = 3 \times \left(1 - \frac{0.8^2}{1^2}\right)^{-1/2} =$$

Observer B records the torch to be on for 5 s.

Test of time dilation: muon decay

Particles called **muons** created in the upper atmosphere travel to Earth at speeds extremely close to the speed of light, 0.9994c. However, muons are unstable particles that decay on their passage towards the Earth. By the time muons reach the Earth, most should have decayed – but this is not observed.

The proper time, t_0, is the time measured in the reference frame of the muon. The proper time represents the average lifetime (or rest half-life), ~2.2 µs. Measurements made of their lifetimes using clocks on Earth are subject to time dilation. On Earth their lifetime is calculated to be

$$t = t_0 \left(1 - \frac{v^2}{c^2}\right)^{-1/2} = 2.2 \times \left(1 - \frac{0.9994^2}{1^2}\right)^{-1/2} = 64\,\mu s$$

i.e. over 28 rest half-lives. The actual number of muons reaching the Earth is observed to be consistent with this result within experimental error, and this provides evidence for the theory of relativity.

Length contraction

Measuring the length of an object in a stationary frame is achieved by determining the difference between the positions of the end points. When an object is moving, the positions of the end points must be recorded simultaneously. Simultaneity is relative, so length is relative. The length l_0 of an object measured in the rest frame of the object is called its **proper length** or **rest length**. Measurements of the length in any other inertial frame that is moving with a velocity v parallel to the length of the object is always less than the proper length. The equation is similar to that for time dilation (page 215) but the proper length l_0 is *divided* by the relativity factor γ:

$$l = \frac{l_0}{\gamma} = l_0 \left(1 - \frac{v^2}{c^2}\right)^{1/2}$$

This is referred to as **length contraction**.

Example

Observer A, on a moving train travelling at a uniform speed of 0.8c, measures the length of the carriage to be 15 m. Determine the length of the carriage for an observer B standing on the platform as the train moves through the station.

Answer

Using the expression above, we obtain

$$l = l_0 \left(1 - \frac{v^2}{c^2}\right)^{1/2} = 15 \times \left(1 - \frac{0.8^2}{1^2}\right)^{1/2} = 9\,m$$

So the carriage according to B is only 9 m long.

Essential Notes

Muons are unstable subatomic particles. Once created, they decay (into other particles) in a very short time. The lifetime of a muon is the time interval between its creation and its decay in its own reference frame, and its average lifetime (rest half-life) is 2.2 µs (2.2×10^{-6} s).

Essential Notes

The Earth is regarded to a good approximation as an inertial frame, i.e. moving with constant speed.

Mass and energy

One other consequence of relativity is the effect it has on mass. The faster an object moves, the more massive it becomes. An object in a stationary frame has a mass m_0, called its **rest mass**. Its mass when it is moving with a velocity v is given by another of the relativity equations:

$$m = m_0 \left(1 - \frac{v^2}{c^2}\right)^{-1/2} = \gamma m_0$$

This is often referred to as its **relativistic mass**. An object that increases its speed also increases its kinetic energy and simultaneously its mass; the effects become significant as v approaches c. In fact, if v were to equal c, the mass would become infinite, so in practice no object with mass can move at a speed equal to or greater than the speed of light.

Mass–energy

A further and more profound consequence of Einstein's theory of relativity is that mass can be considered to be another form of energy. The law of the conservation of energy becomes the law of the conservation of mass–energy. An object that has a mass m has an equivalent energy E given by Einstein's most famous equation:

$$E = mc^2$$

For an object at rest with rest mass m_0,

$$E_0 = m_0 c^2$$

where E_0 is called its **rest energy**.

Total energy

If an object is moving, it will possess not only a rest energy but also a kinetic energy due to its motion, and the total energy is the sum of its rest energy and its kinetic energy (E_k):

$$E = E_0 + E_k = m_0 c^2 + E_k$$

which can be written in the form

$$E = m_0 c^2 \left(1 - \frac{v^2}{c^2}\right)^{-1/2} = \gamma m_0 c^2$$

This is the same as putting the relativistic mass into the equation $E = mc^2$.

Essential Notes

Einstein's mass–energy equation was used when calculating nuclear binding energies in the core of Unit 5, page 22.

Essential Notes

The familiar expression for the kinetic energy, $E_k = \frac{1}{2}mv^2$, is a classical equation that is a good approximation when the velocity is well below the speed of light, c. When an object travels at relativistic speed, its kinetic energy is $(m - m_0)c^2$. It can be shown that this expression reduces to $\frac{1}{2}mv^2$ when $v \ll c$.

Example

What is the rest energy (in MeV) of a muon if its rest mass is 1.88×10^{-28} kg? What would be the kinetic energy (in MeV) of a muon travelling at a speed of $0.9994c$?

Answer

Converting mass in kg to atomic mass units gives

$$\text{mass}\,(m_0) = \frac{1.88 \times 10^{-28}}{1.66 \times 10^{-27}} = 0.1133\,\text{u}$$

Rest energy $E_0 = m_0 c^2 = 0.1133\,\text{u} \times 931.3\ \text{MeV u}^{-1} = 105.5\ \text{MeV}$

$$\text{Total energy } E = m_0 c^2 \left(1 - \frac{v^2}{c^2}\right)^{-1/2} = 105.5 \left(1 - \frac{0.9994^2}{1^2}\right)^{-1/2} = 3046\ \text{MeV}$$

The muon's kinetic energy is $E - E_0 = 3046 - 105.5 = 2940\ \text{MeV}$.

Examiners' Notes

The expressions for time dilation, length contraction and energy are given in the Data and Formulae Booklet.

Essential Notes

For conversions between units of mass and energy see the core of Unit 5, page 23.

Examiners' Notes

Don't forget to subtract the rest energy of the particle when calculating the kinetic energy.

Practice exam-style questions

1 The diagram shows incident radiation falling on a metal surface, which gives rise to the photoelectric effect.

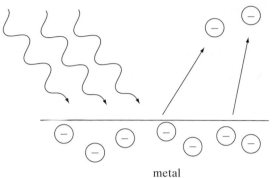

(a) Explain briefly the photoelectric effect as described by Einstein.

_____

_____

_____

_____

_____

_____ 3 marks

(b) The metal sodium has a work function value of 2.3 eV. Determine the threshold frequency.
($h = 6.63 \times 10^{-34}\,\text{J s}$, $1\,\text{eV} = 1.60 \times 10^{-19}\,\text{J}$)

_____

_____ 2 marks

(c) Sodium is irradiated with light of wavelength $4.6 \times 10^{-7}\,\text{m}$. Calculate the maximum velocity of the photoelectrons emitted. (Electron mass $m = 9.11 \times 10^{-31}\,\text{kg}$, $c = 3.00 \times 10^{8}\,\text{m s}^{-1}$)

_____

_____

_____

_____

_____

_____ 4 marks

(d) Determine the stopping potential, V_s.

_____

_____

_____ 2 marks

Total Marks: 11

2 An energetic proton in a cosmic ray shower is recorded to have a kinetic energy of 2.5×10^9 eV.

(a) Calculate the total energy (in MeV) of the proton. (Proton rest mass $m_0 = 1.67 \times 10^{-27}$ kg $= 1$ u, $c^2 = 931.3$ MeV u^{-1})

_____

_____

_____

_____ 2 marks

(b) Hence, using your answer to (a), determine the speed of the proton relative to the ground.

_____

_____

_____

_____

_____

_____

_____ 3 marks

Total Marks: 5

3 A beam of electrons with a speed v is directed into a uniform magnetic field of flux density B in a direction perpendicular to the field lines. The electrons move in a circular orbit in the field with a radius r. ($e = 1.60 \times 10^{-19}$ C, $e/m = 1.76 \times 10^{11}$ C kg^{-1})

(a) Explain why the electrons move in a circular orbit, and write the force equation for this motion.

_____

_____

_____

_____ 3 marks

(b) If the speed of the electrons is $2.6 \times 10^7\,\mathrm{m\,s^{-1}}$ and the flux density B is $10\,\mathrm{mT}$, calculate the radius of the orbit.

_____

_____

_____ 2 marks

(c) Show that the time for one orbit is given by the expression

$$T = \frac{2\pi m}{Be}$$

_____

_____

_____

_____ 2 marks

(d) Calculate the time for one orbit.

_____

_____ 2 marks

Total Marks: 9

4 The diagram shows an experimental arrangement, similar to that which Millikan used, to determine the charge of the electron. X and Y are two horizontal plates.

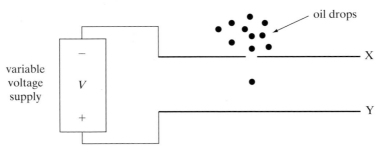

(a) When there is no potential difference between the plates X and Y, the speed of a charged oil drop increases until it is falling at a constant speed. Explain why this happens.

_____

_____

_____

_____

_____ 2 marks

(b) **(i)** An oil drop of mass 1.92×10^{-14} kg is held stationary between the two horizontal plates, which are 20 mm apart. If the potential difference is 2350 V, calculate the charge on the drop. ($g = 9.81$ m s^{-2})

_____

_____

_____

_____ 2 marks

(ii) Suggest what this means about the surplus or deficiency of electrons it carries.

_____

_____ 2 marks

(c) Show that the radius of the oil drop is 1.8×10^{-6} m. (Density of the oil is 800 kg m^{-3}.)

_____

_____

_____

_____ 3 marks

(d) Calculate the terminal velocity of the oil drop when no potential difference is applied. (Viscosity of air between the plates, $\eta = 1.8 \times 10^{-5}$ N s m^{-2})

_____

_____

_____

_____

_____ 3 marks

Total Marks: 12

Answers, explanation, hints and tips

Question	Answer		Marks
1 (a)	The energy of an incident photon *(hf)* is given totally to a surface electron, releasing it as a photoelectron.	(1)	
	The minimum amount of energy needed to release a photoelectron is called the work function, Φ. If $hf < \Phi$, then no photoelectrons are emitted. The threshold frequency is given by $hf_0 = \Phi$.	(1)	
	The kinetic energy of the photoelectrons ranges from zero to a maximum value depending on the number of electron collisions before it leaves the surface.	(1)	3
1 (b)	Threshold frequency is given by		
	$$f_0 = \frac{\Phi}{h}$$	(1)	
	$$f_0 = \frac{2.3 \times 1.60 \times 10^{-19}}{6.63 \times 10^{-34}} = 5.6 \times 10^{14} \text{ Hz}$$	(1)	2
1 (c)	Using Einstein's photoelectric equation		
	$$\frac{1}{2}mv^2_{\text{max}} = hf - \Phi = h\frac{c}{\lambda} - \Phi$$	(1)	
	$$\frac{1}{2}mv^2_{\text{max}} = \left(6.63 \times 10^{-34} \times \frac{3.00 \times 10^8}{4.6 \times 10^{-7}} \right) - (2.3 \times 1.60 \times 10^{-19})$$		
	$$= 6.44 \times 10^{-20}$$	(1)	
	Therefore		
	$$v = \sqrt{\frac{2(hf - \Phi)}{m}} = \sqrt{\frac{2 \times 6.44 \times 10^{-20}}{9.11 \times 10^{-31}}}$$	(1)	
	$$= 3.8 \times 10^5 \text{ m s}^{-1}$$	(1)	4
1 (d)	Using $\frac{1}{2}mv^2_{\text{max}} = eV_\text{S}$	(1)	
	$$V_\text{S} = \frac{6.44 \times 10^{-20}}{1.60 \times 10^{-19}} = 0.40 \text{ V}$$	(1)	2
			Total 11
2 (a)	$$m_0c^2 = 1.67 \times 10^{-27} \times (3.00 \times 10^8)^2$$		
	$$= 1.50 \times 10^{-10} \text{ J} = \frac{1.50 \times 10^{-10}}{1.60 \times 10^{-13}} \text{MeV} = 939 \text{ MeV}$$	(1)	
	$$E = m_0c^2 + E_\text{K} = 939 + 2500 = 3440 \text{ MeV}$$	(1)	2

Question	Answer		Marks
2 (b)	$E = \dfrac{m_0 c^2}{\sqrt{1 - v^2/c^2}}$		
	$\sqrt{1 - \dfrac{v^2}{c^2}} = \dfrac{m_0 c^2}{E} = \dfrac{939}{3440}$		
	$1 - \dfrac{v^2}{c^2} = \left(\dfrac{939}{3440}\right)^2 = 7.45 \times 10^{-2}$	(1)	
	$\dfrac{v^2}{c^2} = 1 - (7.45 \times 10^{-2}) = 0.925$		
	$\dfrac{v}{c} = \sqrt{0.925} = 0.962$	(1)	
	$v = 0.962 \times 3.00 \times 10^8 = 2.89 \times 10^8 \text{ m s}^{-1}$	(1)	3
			Total 5
3 (a)	At any point, the electron is acted upon by a magnetic force, F, that is perpendicular to the electron's velocity, v.	(1)	
	The force causes a centripetal acceleration, a, which always acts towards the centre of curvature of a circular path. Since the magnetic field is uniform, the orbital path is a complete circle.	(1)	
	From Newton's second law, $F = ma$, where m is the mass of an electron,		
	$Bev = \dfrac{mv^2}{r}$	(1)	3
3 (b)	Rearranging		
	$r = \dfrac{mv}{Be} = \dfrac{v}{B}\dfrac{m}{e} = \dfrac{2.6 \times 10^7}{10 \times 10^{-3} \times 1.76 \times 10^{11}}$	(1)	
	$= 0.0148 \text{ m} = 14.8 \text{ mm}$	(1)	2
3 (c)	One complete orbit is the circumference of a circle of radius r, i.e. $2\pi r$, and if the electrons are travelling at a speed v, then		
	time $T = \dfrac{\text{distance}}{\text{velocity}} = \dfrac{2\pi r}{v}$	(1)	
	So		
	$T = \dfrac{2\pi}{v} \times \dfrac{mv}{Be} = \dfrac{2\pi m}{Be}$	(1)	2
3 (d)	$T = \dfrac{2\pi}{10 \times 10^{-3} \times 1.76 \times 10^{11}}$	(1)	
	$= 3.6 \times 10^{-9} \text{ s}$	(1)	2
			Total 9
4 (a)	Initially, the downward force due to gravity (weight) is greater than the upward drag force, and so the oil drop accelerates downwards.	(1)	
	Drag forces increase with velocity, and when terminal velocity (constant speed) is reached, the downward force (the oil drop's weight) is equal to the upward drag force.	(1)	2
4 (b) (i)	$Q = \dfrac{mgd}{V} = \dfrac{1.92 \times 10^{-14} \times 9.81 \times 2.0 \times 10^{-2}}{2.35 \times 10^3}$	(1)	
	$= 1.6 \times 10^{-18} \text{ C}$	(1)	2

Question	Answer		Marks
4 (b) (ll)	Charge on oil drop is positive because of polarity of supply shown on the diagram.	(1)	
	The oil drop lacks 10 electrons.	(1)	2
4 (c)	Using $m = \dfrac{4}{3}\pi r^3 \rho$ and rearranging gives		
	$r = \sqrt[3]{\dfrac{3m}{4\pi\rho}}$	(1)	
	$= \sqrt[3]{\dfrac{3 \times 1.92 \times 10^{-14}}{4 \times \pi \times 800}}$	(1)	
	$= 1.8 \times 10^{-6}$ m	(1)	3
4 (d)	$6\pi\eta r v = mg$		
	$v = \dfrac{mg}{6\pi\eta r}$	(1)	
	$v = \dfrac{1.92 \times 10^{-14} \times 9.81}{6\pi \times 1.8 \times 10^{-5} \times 1.8 \times 10^{-6}}$	(1)	
	$= 3.1 \times 10^{-4}$ m s^{-1}	(1)	3
			Total 12

Glossary

absolute magnitude the apparent magnitude a star would have if it were placed at a standard distance of 10 parsec from the Earth

absolute temperature scale temperature scale measured in kelvin (K), which has 0 K as absolute zero; $0\,K = -273.15\,°C$; the other defining point on the scale is the triple point of water

absolute zero the lowest possible temperature: the temperature at which there is zero kinetic energy of the particles (particles are stationary except for quantum-mechanical motion)

absorption spectrum dark lines on an otherwise continuous bright spectrum

accommodation the ability of the eye to change its focal length by changing the shape of the lens

action potential the pattern of changing potential difference that is transmitted down a nerve cell

active galaxy a highly energetic galaxy, possibly containing a supermassive black hole

activity the activity of a radioactive source is the number of emissions per second; measured in becquerels (Bq); $1\,Bq = 1$ emission per second

adiabatic process a thermodynamic process during which no heat enters or leaves; $Q = 0$

Airy disc the bright central region in an optical diffraction pattern caused by light entering a circular aperture

alpha (α) radiation short-range, highly ionising radiation consisting of helium nuclei

alpha (α) particle particle formed from two protons and two neutrons (a helium nucleus); emitted by the nuclei of some radioisotopes

angular acceleration the rate of change of angular velocity, given by

$$\alpha = \frac{\Delta\omega}{\Delta t}$$

angular displacement a change in angular position, given by $\Delta\theta = \theta_2 - \theta_1$

angular impulse a change in angular momentum

angular magnification the magnifying power of a refracting telescope, given by the ratio of the objective focal length to the eyepiece focal length

angular momentum a vector quantity in rotational motion, given by $L = I\omega$

angular resolution a measure of the ability of a telescope to distinguish between adjacent astronomical features or objects (also called the resolving power)

angular velocity the rate of change of angular displacement with respect to time, given by

$$\omega = \frac{\Delta\theta}{\Delta t}$$

apparent magnitude the apparent brightness of a star expressed on the magnitude scale

A-scan (amplitude scan) simple echo technique used in ultrasonic measurements; the size and time of a reflected pulse gives the position of the change in the medium

astigmatism a focusing problem in the eye caused by a cornea that curves more in some directions than in others

Astronomical Unit (AU) the average distance between the Earth and the Sun: $1.49 \times 10^8\,km$

atomic mass unit (u) unit of mass defined as 1/12 of the mass of a carbon -12 atom

atrium (plural atria) the smaller chamber(s) of the heart which pump(s) blood into the ventricle(s)

attenuation the reduction of intensity as a wave travels through a material

audiogram a graph showing hearing loss as a function of sound frequency

audiometer a device used to test hearing by producing sounds of known intensity level over a range of frequencies

Avogadro constant (N_A) the number of particles in a mole of a substance; $N_A = 6.02 \times 10^{23}$

Avogadro's law law stating that equal volumes of gases, at the same temperature and pressure, contain the same number of molecules

axis of rotation an axis around which a rigid body rotates

background radiation count rate the number of emissions per second that are detected due to radiation from the environment

Balmer series a series of emission or absorption lines in the visible spectrum of hydrogen

beta (β) radiation ionising radiation in the form of fast-moving electrons (or positrons)

Big Bang the explosion event approximately 14 billion years ago that cosmologists consider the beginning of the Universe

binary star system two stars revolving about a common centre of mass

binding energy (nuclear) the energy required to separate a nucleus into its constituent protons and neutrons

binding energy per nucleon the average energy required to remove each proton or neutron from a nucleus

black body an object that absorbs all the thermal radiation that falls upon it and reflects none; it is also a perfect emitter whose radiation depends only on its temperature

black dwarf the end stage of a low-mass star such as the Sun

black hole highly dense matter around which gravity is so strong that the escape velocity exceeds the speed of light

black-body curve the intensity of radiation emitted by a black body as a function of wavelength (or frequency) and characteristic of its temperature

black-body radiation the radiation emitted by a black body or, in practice, any very hot object such as a star

blue shift a decrease in observed wavelength of radiation emitted by an object approaching an observer

Boltzmann constant (k) a constant that links the absolute temperature of a gas to the average kinetic energy of its molecules: k is equal to the ratio of the molar gas constant R to the Avogadro constant N_A

Boyle's law law stating that for a fixed mass of an ideal gas at constant temperature, the pressure of the gas is inversely proportional to its volume: $pV = $ constant

brake power or brake horsepower (bhp) the output power from an engine = output torque × angular velocity

bremsstrahlung the continuous spectrum of X-rays given off by electrons as they are decelerated by a target

sensorineural loss loss of hearing due to damage in the inner ear, auditory nerve or brain

solar constant the average amount of energy received from the Sun per second per square metre on Earth (about $1400 \, \mathrm{W \, m^{-2}}$)

sound intensity level intensity level of a sound relative to the intensity at the threshold of hearing; measured in decibels $= 10 \log_{10} (I/I_0)$

specific acoustic impedance (Z) term used in ultrasound imaging, equal to the product of the medium's density, ρ, and the speed of sound, c; unit $\mathrm{kg \, m^{-2} \, s^{-1}}$

specific charge the ratio of the charge of an elementary particle to its mass

specific heat capacity the energy required to raise the temperature of a 1 kg mass of a substance by 1 K; unit $\mathrm{J \, kg^{-1} \, K^{-1}}$

specific latent heat of fusion the energy needed for 1 kg of a solid to change to a liquid, with no increase in temperature; unit $\mathrm{J \, kg^{-1}}$

specific latent heat of vaporisation the energy needed for 1 kg of a liquid to change to a gas, with no increase in temperature; unit $\mathrm{J \, kg^{-1}}$

spectral class the type of star, as classified by its temperature and hence spectral properties; the temperature sequence of spectral classes is OBAFGKM (Oh Be A Fine Girl, Kiss Me)

spherical aberration the distortion of an image due to imperfections in the mirror or lens causing differing focal lengths

standard candle an astronomical object of known intrinsic brightness, for example a supernova, that is used to determine astronomical distances

Stefan's law relation that gives the total energy emitted per square metre per second from an object at a given temperature T to be proportional to T^4

stellar spectroscopy the analysis of spectra from stars in order to obtain information about temperature, composition, etc.

Stokes' law a law giving the frictional force on a spherical ball moving through a viscous medium

stopping potential the voltage required to stop the emission of photoelectrons from the surface of a metal

supergiant highly luminous star with masses 10 to 100 times that of the Sun and high core temperatures

supermassive black hole black hole having a mass of 10^6 to 10^9 times that of the Sun, usually found at the centres of galaxies

supernova explosive death of a star, caused by sudden onset of nuclear burning or energetic shock wave; one of the most energetic events in the Universe

synchrotron radiation radiation emitted by charged particles as they accelerate in a strong magnetic field

thermal efficiency (ε) the ratio of the indicated power to the input power for an engine; also given by

$$\varepsilon = 1 - \frac{Q_{\mathrm{out}}}{Q_{\mathrm{in}}}$$

thermionic emission the emission of electrons, usually in a vacuum, from a heated conductor (cathode)

thermometric property a property of a substance that changes with temperature, such as the volume of a gas, the length of a mercury column or the electrical resistance of a wire; can be used to provide a temperature scale

threshold frequency the minimum frequency of incident electromagnetic radiation for which photoelectrons are liberated from the surface of a material

threshold of hearing (I_0) the lowest sound intensity that can be detected by a healthy human ear, defined as $1 \times 10^{-12} \, \mathrm{W \, m^{-2}}$

time dilation a consequence of special relativity: an elapsed time interval appears to be longer for an observer in a frame of reference moving at (high) speed relative to the frame of reference in which the events occur

torque the moment of a force about a point, given by $T = Fs$; unit N m

tracer a radioisotope, usually a gamma emitter, used to track the passage of a substance; for example, radioactive iodine is used in the body to monitor thyroid function, and radioactive gold is used to study coastal erosion

transmission electron microscope (TEM) a microscope based on the transmission of electrons through ultra-thin specimens

transverse wave a wave that vibrates in a direction perpendicular to the direction of travel, e.g. an electromagnetic wave

triple point of water the unique combination of temperature and pressure at which water exists as ice, liquid and water vapour in stable equilibrium; occurs at a temperature of 273.16 K (0.01 °C) and a pressure of 611 Pa (0.006 × normal atmospheric pressure)

Type I supernova supernova explosion caused by a white dwarf star gaining material from a binary companion and reaching a critical mass

Type II supernova supernova explosion caused by the collapse of a massive star when nuclear burning ceases in its core

ultrasound sound waves with a frequency above 20 kHz

unsharpness the lack of quality of an X-ray image (blurring), due to a region of partial shadow

ventricle chamber in the heart which collects blood from an atrium and pumps it out of the heart

virtual image an image caused by rays that do not converge; the image can be seen by the eye but not formed on a screen

viscosity the resistance a fluid has to flow; unit Pa s or $\mathrm{N \, s \, m^{-2}}$

visual acuity the ability of the eye to resolve separate images

wave–particle duality the concept that all energy (and matter) exhibits both wave-like and particle-like properties

wavefront in wave motion, a surface that is tangential to all secondary wavelets based on Huygens' principle

white dwarf a low-mass small star (about Earth-sized) that has exhausted all its nuclear fuel but has a high surface temperature

Wien's displacement law relationship between temperature T of a black body and the peak wavelength λ_{max} (the wavelength at which the emitted radiation intensity peaks): $\lambda_{\mathrm{max}} T = \text{constant} = 2.9 \times 10^{-3} \, \mathrm{m \, K}$

work done the product of the force and the distance moved; for rotational motion of an object it is given by $W = Fs\theta = T\theta$; for a gas at constant pressure it is given by $W = p \Delta V$

work function the minimum energy required to remove a photoelectron from the surface of a metal

Index

Notes

Notes

Notes